The Calendar Code

The Calendar Code

The Calendar Based on 0 through 6

David Braverman

Library of Congress Control Number: 2009910095
ISBN: Hardcover   978-1-4836-3064-9
                   Softcover   978-1-4836-3063-2
                   Ebook   978-1-4836-3065-6

This book was printed in the United States of America.

Rev. date: 04/20/2013

To order additional copies of this book, contact:
Xlibris Corporation
1-888-795-4274
www.Xlibris.com
Orders@Xlibris.com
68655

CONTENTS

Tables 7-9

The Modulo 7 Clock 10

Preface 11

A Brief History of the Gregorian Calendar 13

Preliminary review of addition of signed numbers 14

Preliminary review of solving simple first degree equations 18

The Mission 20

What is a Modulo System? 22

Getting Around on the Modulo 7 Clock 25

REDUCING positive numbers in Modulo 7 27

Negative Numbers and Negative Equivalents in Modulo 7 31

Reducing a Negative Number in Modulo 7 35

Reducing in modulo 7 as you work 38

Solving equations in modulo 7 41

The Basics of the Gregorian calendar 44

The Main Number for Years is 5 47

The Mathematics of the Calendar 50

Century Codes 51

The Year Codes: The messy part 53

Finding Year Codes Locally by Counting 55

Finding Year Codes Using the Year Code Algorithm 57

The Codes of The Months and Their Origins 61

Friday the 13th 64

Task A: Solving The Birthday Problem 66

Task A: Practice in Finding the Day of the Week of a Date
          or the Birthday Problem 70

Task A: Another test.. 72

Task B: Find dates in a *given month* in a *given year* that fall on a
         given day of the week.or *The Appointment Calendar Problem* 74

Task C: Finding YEARS with a Calendar Identical to a given year
         or re-using calendars.. 80

Task D: The Law & Order Problem.. 86

Task E: Find the month(s) in which a given date in a given year
         can fall on a given day of the week.. 90

Task F: *Reducing Large Numbers (3 and 4 digits) in Modulo 7* 97

Final examination for Test A: 102

Answers to TEST A:... 103

Final examination for Test B: The Appointment Calendar Problem 112

Answers to Test B: The Appointment Calendar Problem........................... 113

Final examination for Test C: The Law & Order Problem 119

Answers to Test C: The Law & Order Problem... 120

Final examination for Test D: Finding future years
         to re-use the calendar... 126

Answers to Test D: Finding future years to RE-USE the calendar 127

Final examination: Test E. 131

Test E: Answers 132

Final examination: Test F: *Reducing Large Numbers in Modulo 7* 135

Answers to Test F: Reduce in modulo 7. No shortcuts were taken.
         That is your job. 136

Cumulative examination 138

Answers to the cumulative examination 139

Afterword 143

Index.. 144

Multiples of 7

7 x 1 = 7	7 x 11 = 77
7 x 2 = 14	7 x 12 = 84
7 x 3 = 21	7 x 13 = 91
7 x 4 = 28	7 x 14 = 98
7 x 5 = 35	7 x 15 = 105
7 x 6 = 42	7 x 16 = 112
7 x 7 = 49	7 x 17 = 119
7 x 8 = 56	7 x 18 = 126
7 x 9 = 63	7 x 19 = 133
7 x 10 = 70	7 x 20 = 140

Multiples of 4 (for leap years)

There are 24 leap years (divisible by 4) in a *normal* century. *Leap centuries (divisible by 400)* have *25 leap years*. Years ending in 00 are *centennial years*, an extra leap year.

Only the *last two digits* are shown to help you find leap years.

This is explained in the book as are all aspects of the system.

04	52
08	56
12	60
16	64
20	68
24	72
28	76
32	80
36	84
40	88
44	92
48	96

The codes for months

January	3	April	2	July	2	October	3
February	6	*May	4	*August	5	November	6
March	6	*June	0	September	1	December	1

* These months have a unique code. There is no other month with this code.

Memorize the month codes in groups of three: 3 6 6   2 4 0   2 5 1   3 6 1

The codes for select *non-leap* years:

1930, 1941, 1947, 1958, 1969, 1975, 1986, 1997, 2003, 2014, 2025 **have code 0**
They are useful for counting up to subsequent years

DUAL Codes for select leap years: Read the chapter on The Year Codes.
Leap years with same dual codes are separated by 28 years.
    The *first code* is used in January and February.
    The *second code* is used from March to December.

1936, 1964, 1992, 2020 have the dual codes (0, 1)

1948, 1976, 2004, 2032 have the dual codes (1, 2)

1932, 1960, 1988, 2016 have the dual codes (2, 3)

1944, 1972, **2000, 2028 have the dual codes (3, 4)

1956, 1984, 2012, 2040 have the dual codes (4, 5)

1940, 1968, 1996, 2024 have the dual codes (5, 6)

1952, 1980, 2008, 2036 have the dual codes (6, 0)

**2000 is a centennial year. This occurs every 400 years.

The century codes 0, 5, 4 and 2

16th century (partial) code 5   17th century code 4   18th century code 2

19th century code 0   20th century code 5   21st century code 4

22nd century code 2   23rd century code 0   24th century code 5

The Modulo 7 Clock

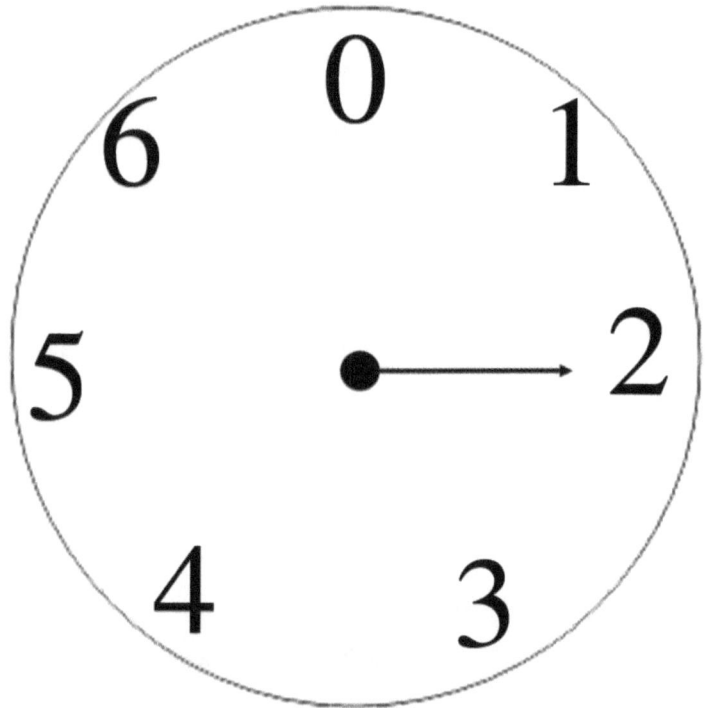

Preface

I am a retired high school mathematics teacher. My Bachelor of Science degree in mathematics was earned at the State University of New York at Stony Brook in 1967. My Master of Arts degree in mathematics was earned at Lehman College of the City University of New York. I started teaching in 1967 at the age of 22 but my insatiable curiosity of numbers started in my early adulthood.

My aunt was an elementary school principal and she taught me a way to tell the day of the week of any date in the *current* year. It was severely *flawed*. I was unable to handle dates in January or February. That was because the current year was a *leap year*.
When I learned more mathematics, I taught myself how to handle dates in January and February in *leap years*. I also unraveled why there were only *24 leap years* in a normal century. I taught myself about the elusive centennial year that occurs every 400 years. In short, I refined the system to its current state. These topics are discussed in the Brief History of the Gregorian calendar.

The system is as complete as I could make it. All the holes in my aunt's technique have been filled in to the best of my ability. The problems in the book are all solved in detail. Exercises are provided for students to solve. Cover the answers with your hand or just don't look at them. You don't want to be distracted. The answers will be there when you are done.

There are six tasks for you to learn based on the calendar.
A date has *four variables*: the *month*, the numerical *date*, the *year* and the *day of the week* on which the date falls. You will learn the *date algorithm*

that enables you to solve these problems. In *four* of the tasks, *one* of these variables is omitted. There is a test for each task. The answers follow each test and are shown in detail with explanations.

There is a review of adding signed numbers and solving simple equations. If you have these skills already, start at the chapter **What is a Modulo System?**

The twelve codes for the months are *derived in detail*. *Memorize* these codes. Any one of the five tasks is more impressive if it is done *mentally*. However, writing the twelve codes on a scrap of paper will not diminish the mystery of how it is done.

Over the years, people, from my students to my dental hygienist have asked me, *"how do you do that?"* and sometimes I supplied them with a few sheets of instruction explaining the system. But that was never enough. Thus, a book. What I know, you will know.

I believe that anyone who loves math as much as I will enjoy this trip through time and share it with others. Tell your friends the day of the week on which they were born. Learn other tricks you can perform with your mind. Be a human calendar.

Do not rush through this. Take your *time*.

David Braverman December 3, 2009

A Brief History of the Gregorian Calendar

On *February 24, 1582*, Pope Gregory XIII declared that the *Gregorian calendar*, which is in use today, will be the *official* calendar, replacing the Julian calendar.

On Tuesday, October 5, 1582, the calendar was adopted by some Catholic countries. France began using it on Friday, December 10, 1582. Britain began in 1752 and some Asian countries still used the Julian calendar as late as the twentieth century.

The length of the year in the Gregorian calendar is 365.2425 days. That is *one four-hundredth* of a day more than 365.24 days.

100 x .2425 = 24.25 so in 100 years, there would *24 extra days we call LEAP DAYS*. (We don't handle *partial* days, so .25 of a day or *one-fourth of a day* lost each day is left unresolved.)

The Pope decreed that the leap day would be February 29. The years in which that occurred would be *divisible by four,* as they were in the Julian calendar.

Thus, there would be 24 leap years in a century. The big change from the Julian calendar was the correction of the one-fourth day lost *every 400 years*. Every *400* years there would be a *special leap* year called a *centennial year*. 1600 and 2000 are the only centennial years so far. The next centennial year is 2400. A simple way to recognize a centennial year is to see if the *first two digits* of the year is a multiple of (or divisible by) 4.

Today, scientists use an atomic (cesium) clock to make adjustments in *leap seconds* for some years due to irregularities in the daily rotation of the Earth or in its revolutions about the sun. We will *NOT* be delving into that area.

Preliminary review of addition
of signed numbers

To understand signed numbers, you must understand the number line.

```
<--|-----|-----|-----|-----|-----|-----|-----|-----|-----|------|-----|-----|-----|-----|-----|-----|-----|-----|->
   -9   -8   -7   -6   -5   -4   -3   -2   -1    0   +1   +2   +3   +4   +5   +6   +7   +8   +9
```

The arrows at the ends indicate that the line is infinite. Numbers get larger as you move to the right and smaller as you move to the left. Forever. The concept of *absolute value* is *essential* to understanding the addition of signed numbers. First off all *get rid of the elementary school concept of addition as building up* numbers to a sum that is greater than either of the numbers being added. The result of adding two signed numbers might very well be *smaller* than *either* of them.

The absolute value of a number is its *distance from 0* on the number line. For example, the absolute value of -6 is 6 because it is 6 spaces to the left of 0. +6 also has an absolute value of 6. It is just in another position, *to the right* of zero. Most people think of absolute value as *"the number without the sign"* which it *is*. That is correct but not very mathematical. Use it if you wish.

The words *combine and add are* interchangeable. The result will be designated as the *sum*. All of these words will be used to indicate addition of signed numbers.

Numbers being added are written left to right, *each number with a sign* except for the first number which does need a sign *if it is positive*. The numbers form a chain. This chain signifies addition.

+9 − 12 + 8 − 9 − 15 is what such a chain looks like. Note that the *signs are links* in the chain. That is why *you can not omit a sign* (except for the first number as noted before). When adding (or combining) *two* signed numbers, two cases may be encountered.

Case I. The numbers have the *same* signs. In this case, the absolute values are *irrelevant!* The numbers are *added as you learned in elementary school.* The sign of the sum is the *same as the sign of the numbers.* Think of them as belonging to the same team. Whatever team they both play for, their sum will be on the same team.
One sign, one team.
For example, -7 - 13 both have a negative sign. Add as you did in elementary school. Their *sum* is -20. In the example -2 − 5 − 3, add as you did in elementary school. Just don't forget the sign of the answer is negative!
For example, -2 − 5 − 3 = -10.
One team, one sign.

Case II. The signs of the two numbers are different. This is the more difficult case.
In Congress, when the majority is Republican, Republicans dominate.
In signed number addition, the same effect occurs.
The *larger absolute value dominates.* If the numbers are opposite in sign and have the same absolute value, their sum is 0.

Another analogy you can use is a scale (like the scales of justice) that weighs a negative number against a positive number. The weights are the absolute values. Whichever one tips the scale has the dominant "weight" and its sign is the sign of the sum. If the scale balances, the sum is 0.

The sum of two *opposites (also known as complements, or additive inverses)* is 0. Neither number dominates. They "weigh" the same. For example, the sum of -5 and +5 is 0.

If the numbers are *not* opposites, decide which number has the larger absolute value. The sign of that number will be the sign of the result.

Then <u>subtract</u> the absolute values. If you prefer the scale analogy, find the difference of the weights of the numbers.

When the signs of *two numbers* are *different*, think of the word *difference*. When you associate the words *different* and *difference*, adding will be easier.

In the case of +8 – 27, the larger absolute value belongs to -27, or -27 "outweighs" +8, and that makes the sign of the result negative. Then simply subtract 27 – 8 = 19.

The sum is -19.

Resist the temptation to say things like "-27 is larger than +8" because in fact, -27 is *smaller* than +8 on the number line because - 27 is to the *left* of +8.

Example: -8 – 3 + 11 – 12 + 3

Do you see the two opposites in the problem? -3 and +3 have a sum of 0. *Cross out <u>both</u> -3 and +3* from the problem. The problem is now -8 + 11 – 12.

Method 1. First find numbers with the *same sign* and add them.

One team, one sign.

The positive "team" sum is +11 because that is the *only positive number*.

The negative "team" has a sum of –20 because -8 – 12 = -20.

Now the problem is +11 – 20 and the signs are *different*. Find the *dominant sign*. The *negative number has the larger absolute value* so the result will be *negative*. The difference of 11 and 20 is 9. Then the answer is -9.

Thus, -8 – 3 + 11 – 12 + 3 = -9.

Method 2. Another way to add these numbers is *from left to right*, two at a time. This may be faster for those who are able to add quickly.

-8 – 3 = -11 → -11 + 11 = 0 → * 0 – 12 = -12 → -12 + 3 = -9

*It is not necessary to write 0.

Exercises. Answers follow.

1) Find the sum of 10 and -8.
2) Combine -10 + 6 – 4 – 3 + 13

3) -6 + 14 – 9

4) -7 + 4 + 3

5) 12 – 19 + 6

Answers:

1) Find the sum of 10 and -8.

   The signs are different. The positive number has the *larger absolute value.*
   The difference of 10 and 8 is 2, *thus* the sum is +2 or 2.

2) Combine -10 + 6 – 4 – 3 + 13

   The negative numbers add up to -17. The positive numbers add up to +19.
   The problem becomes -17 + 19.
   The absolute value of the positive number is larger than that of the negative
   number.
   The *difference* of 19 and 17 is 2. The result is +2 or simply 2.

3) -6 + 14 – 9

   The sum of the negative numbers is -15.
   The problem becomes -15 + 14. There is only one positive number, +14.
   The *absolute value of the negative is larger* than that of the positive number.
   The *difference* of 15 and 14 is 1 thus, the sum is -1.

4) -7 + 4 + 3

   The sum of the positive numbers is +4 + 3 = +7 or 7.
   The problem becomes -7 + 7 = 0.

5) 12 – 19 + 6

   The sum of the positive numbers is 12 + 6 = 18.
   The problem becomes 18 – 19. There is one negative number, -19.
   The absolute value of the negative number is larger than that of the positive
   number.
   The *difference* of 19 and 18 is 1 thus, the sum is -1.

Preliminary review of solving simple first degree equations

You will be solving equations in the form $x + a + b = c$, where a, b and c represent signed numbers.

The first step is to add (or combine) the numbers *on the left side of the* = *sign*.

The second step is to *transpose* the signed number on the left, by placing its opposite (additive inverse) on the *right side*.

For example: $x - 5 + 3 = 4$

On the *left side* of the = sign you see $x - 5 + 3$.

This becomes $x - 2$ after adding (or combining) -5 and +3.

Now the equation is $x - 2 = 4$. Then transpose -2 by writing $x = 4 + 2$.

Now combine on the *right side* of the = sign. Combining: $4 + 2 = 6$.

Then $x = 6$. In this book, if the result is positive, the + sign will be omitted.

The three steps are 1) Combine on the left 2) Transpose 3) Combine on the right

Solve: Answers follow.

1) $x - 4 + 1 = 3$

2) $x + 6 + 4 = 11$

3) $x + 3 + 5 = 4$

4) $x + 6 + 8 = 0$

5) $x - 5 + 3 = 6$

Solutions:

1) x - 4 + 1 = 3
   Add on the LEFT side we get x - 3 = 3 → Transposing: x = 3 + 3
   Solution $x = 6$.

2) x + 6 + 4 = 11
   Add on the LEFT side we get x + 10 = 11 → Transposing: x = 11 − 10
   Solution x = 1.

3) x + 3 + 5 = 4
   Add on the LEFT side we get x + 8 = 4 → Transposing: x = 4 − 8
   Solution $x = -4$.

4) x + 6 + 8 = 0
   Add on the LEFT side we get x + 14 = 0 → Transposing: x = 0 − 14
   Solution $x = -14$

5) x - 5 + 3 = 6
   Add on the LEFT side we get x - 2 = 6 → Transposing: x = 6 + 2
   Solution $x = 8$.

The Mission

The purpose of this book is to empower the reader to be able to perform six tasks listed below. The *prerequisite* is a <u>firm</u> understanding of *signed number addition* and the knowledge of first degree *equations in the form* $x + a + b = c$ (where a, b and c represent positive integers). It will also be necessary to know <u>*multiples of seven*</u> (because there are seven days in a week). Refer to the table of multiples of seven when necessary.

Also, you should be able to *identify* a multiple of four (for leap years). Multiples of four are also on a table.

The author uses the terms subtrahend and minuend to describe subtraction. The *minuend* is the first number (and *usually* the *larger* number). The *subtrahend* is the number that is being subtracted, and it is *usually* smaller than the minuend. This will be useful later on to communicate some principles.

You are urged to learn the *numerical* sequence of the days of the week. Each day appears on a circular clock that represents a system called modulo 7 which has 7 numbers on it, each *representing* a day of the week. There are an *infinite* number of such modulo systems. Modulo 24 is familiar to those in the military. THAT clock is *real* and it has a population (numbers) of 24. The *modulo 7* clock has *a seven numbers, from 0 to 6 with 0 in the "midnight" position*. There are *no other numbers* in modulo 7.

There are *no fractions* in modulo 7.

Each number on the clock represents an infinitely long list of numbers to which it is equivalent. For example, 3 is equivalent to 10, 17, 24, 31 etc. We will be talking about equivalences in modulo 7 in depth.

You will learn how to *reduce* numbers (positive and negative) to the seven numbers in modulo 7 that is 0, 1, 2, 3, 4, 5 or 6. For example, 47 reduces

to 5 and -39 reduces to 3 in modulo 7. Reducing will be discussed in detail later. Each day of the week is assigned a value from 0 to 6.
Saturday is assigned the code of 0 and *Sunday is assigned the code of 1.*

Sun $\equiv 1$ *Mon* $\equiv 2$ *Tue* $\equiv 3$ *Wed* $\equiv 4$ *Thu* $\equiv 5$ *Fri* $\equiv 6$ *Sat* $\equiv 0$

The symbol $\equiv$ means "is *equivalent to*" and it will be explained later.
The student will be able to:

- ✓ Task A. Find the day of the week of *ANY date (or the Birthday Problem).*
- ✓ Task B. Find dates in a *given month* in a *given year* that fall on a *given day of the week.*
- ✓ Task C. Find years that have *identical calendars to a given year* (so you can reuse calendars).
- ✓ Task D. Solve the *"Law&Order Problem":* Given a month, a date and the day of the week on which an episode falls, find probable years in which it was *made.*
  For example, *Monday, October 19* could occur in 1998 and 2009.
- ✓ Task E. Find the month(s) is a particular year in which a given date can occur.
  For example, in 2009, the 19th could fall on a Thursday *only* in *February, March and November.*
- ✓ Task F. Reducing large numbers in modulo 7.

What is a Modulo System?

This world is awash with modulo systems. The Earth is a modulo 24 system itself. The day begins at *zero hour* (also known as midnight in military circles) and progresses 24 hours until a complete revolution has been made. *After 2300 hours (also known as 11PM in civilian circles), it resets to zero.* That is the gist of a modulo system.

Odometers in cars are familiar to most people. Old cars had odometers that registered 99,999 miles and then *reset to 0* when it reached 100,000 miles. This would be a modulo 100,000 system. So when you see a *very* old car with an odometer reading of 350 miles, a sensible conclusion that the car actually had been driven 100,350 miles could be reached. Today, manufacturers install odometers that can register 999,999 miles, essentially a modulo 1,000,000 odometer.

The *year* is a modulo 12 system. There is no 0 month but that is the only difference between it and modulo 7. After December, the month numbered 12, the year resets with year number 1, January. There is no month numbered 13. These are *not the same as month codes*. They just position the months in the course of the year.

For the purposes of this book, one particular modulo system will be used. It is based on the days of the week which are represented numerically by the codes 1, 2, 3, 4, 5, 6 and 0.
These codes represent respectively, Sunday, Monday, Tuesday, Wednesday, Thursday, Friday and Saturday. A picture of the clock is on the cover and within the book.

After Saturday, day 7, the week resets, that is, it begins repeating. The code of Saturday is 0. The day *after* Saturday, Sunday, is represented by the code 1. When presented with a number larger than 6, a multiple of 7 *must* be *subtracted* from it to *reduce it* to one of the seven numbers in modulo 7. You will learn to think in this system and it will be natural for you to think that the day following 6 is 0, because *Friday is followed by Saturday.*

The day after 6 is 0.

Modulo 7 is fairly easy to learn. It is *based on multiples of 7.* In the world of modulo 7, *all multiples of 7 are considered to be equivalent to 0.* That means 7, 14, 14, 21, 28 etc are all considered to be 0 in modulo 7.We use the symbol ≡ when we *reduce* a number *greater than 6* in modulo 7.

$21 \equiv 0 \pmod 7$ is to be read as "21 is equivalent to 0 in mod 7".

In that light, $14 \equiv 0 \pmod 7$, $21 \equiv 0 \pmod 7$, $28 \equiv 0 \pmod 7$ etc.

0 has many faces. In fact *every whole number* has an *infinite* list of aliases.

1 is equivalent to 8, 15, 22, 36, 43 etc. because they are all 1 more than a *multiple of 7.*

We use the symbol ≡ with the goal of getting a result that is *one of the seven numbers on the modulo 7 clock.* If the result of a calculation is *not* on the modulo 7 clock, we *can* use the usual = sign until the result is one of the numbers on the modulo 7 clock.

Persistence is important.

Finding the largest multiple of 7 that can be subtracted from a number is at the heart of modulo 7 as well other modulo systems. We will use subtraction to reduce positive numbers that are not one of the 7 numbers in modulo 7, namely 1, 2, 3, 4, 5, 6 and 0.

Size in a modulo system is a little strange. Is 6 greater than 0 on the clock? 6 is 1 unit counter-clockwise from 0! That would suggest that 6 is smaller than 0. Can 0 be thought of as larger than 6? Not in the normal sense. *Friday is followed by Saturday. 6 is followed by 0.*

The code of 6 is followed by the code of 0 moving clockwise from 6. With that view, you can say that 6 is less than 0. Also, you can say that 6 is 6 units ahead of 0 if you move counter-clockwise from 6 to 0. if you move When moving on a circular path, the number that you see *ahead* of you can also be viewed as *behind* you!

"1 more than 6" means 7 in conventional mathematics.

In modulo 7, it means $6 + 1 \equiv 7 \equiv 0 \pmod 7$.

In modulo 7, the number that follows 6 is 0.

Is Wednesday before or after Thursday? The answer depends on the week to which you refer. If it is the current week, then the answer Wednesday is behind Thursday. If it is next week, then Wednesday is ahead of Thursday. In modulo systems size does *not* matter. The seven numbers in the system are *trapped* on the clock. Think of a carnival spinner with 0, 1, 2, 3, 4, 5 and 6 on it. When the spinner is spun 100 times or so, it still comes to rest on one of the 7 numbers! The numbers on the clock never get larger as you rotate the spinner. When the term is *larger* is used, it is used in the sense of *a number line* which is infinite and has *direction*. Numbers to the *right* are larger than numbers to the *left*. Circles have no direction.

Remember, there *is no* 7 in mod 7. Adding or subtracting 7 or 14 or 21 etc. is *exactly* the same as adding or subtracting 0 in modulo 7. It has no effect.

$3 - 56 \equiv 3 \pmod 7$ because $56 \equiv 0 \pmod 7$ and $3 - 0 \equiv 3 \pmod 7$.

$63 + 5 \equiv 5 \pmod 7$ because $63 \equiv 0 \pmod 7$ and $0 + 5 \equiv 5 \pmod 7$.

Know the multiples of 7.

If you see a multiple of 7 in a problem, cross it out, sign and all.

If you see any *combination* that *creates* a multiple of 7 in a problem, you should *delete* it.

$x + 9 + 5 \equiv 6 \pmod 7$ quickly becomes $x \equiv 6 \pmod 7$ because

$9 + 5 = 14 \equiv 0 \pmod 7$.

More on this will be discussed later.

Getting Around on the Modulo 7 Clock

Working in Modulo 7, these problems will come up in innumerable cases:

a. What is the largest possible multiple of 7 that can be subtracted from a given number?

b. Subtract that multiple of 7 from the given number. This called reducing.

98 _____ 79 _____ 62 _____ 36 _____ 83 _____ 58 _____ 46 _____ 31 _____

a. Answers: 98   77   56   35   77   56   42   28
b. Answers: 0   2   6   1   5   2   4   3

The transition to modulo 7 is easy. We have a circular clock with 7 numbers on it. At the midnight position of the clock is 0 (where one would *expect* a 7 to appear).

The spinner is a convenient metaphor for modulo 7 (and all modulo systems). There is only *one hand on this clock (as in all modulo clocks) as the book cover* shows.

Create a virtual clock in your mind, just like the one on the cover of this book.
Example: 5 + 6
Set the hand (pointer) to 5. Adding 6 is a *clockwise* movement.
Move the hand 6 "hours" clockwise. From 5, you move to 6, 0, 1, 2, 3 and then stop at 4. Then $5 + 6 \equiv 4 \pmod 7$.
Or add conventionally: $5 + 6 = 11$.
The largest multiple of 7 that can be subtracted from 11 is 7.
Subtracting $11 - 7 = 4$. Then $11 \equiv 4 \pmod 7$.
In either case, the result is 4 (mod 7)

Example: 3 – 5

Point the hand of the clock to 3.

Subtracting 5 is a *counterclockwise* movement.

Move the hand 5 "hours" *counterclockwise*. You move to 2, 1, 0, 6 and stop at 5.

Then $3 - 5 \equiv 5$ (mod 7).

Or, add conventionally: $3 - 5 \equiv -2$. *Then $-2 \equiv 5$ (mod 7) <u>must</u>* be true.

We will delve into negative numbers in modulo 7 soon.

Try these. Make sure your answer is less than 7.

Using multiples of 7 is recommended.

If you use the clock mentally, keep track as you rotate from number to number, like train stations.

1)3 + 4   2)1 + 8   3)4 + 6   4)4 + 21   5)6 + 2   6)2 + 14

The results are, respectively, 1)0 2)2 3)3 4)4 5)1 6)2

Tip: If you come across a problem like 5 + 9 + 2 (mod 7), you can *cross out* 5 + 2 because 5 + 2 = 7 *which is equivalent to 0 (mod 7)*. Only +9 remains and $9 \equiv 2$ (mod 7).

If you got these all correct, then you are capable of proceeding. If not, go over the previous pages until you feel more secure.

REDUCING positive numbers in Modulo 7

In modulo 7, to reduce a positive number you will need to find the *largest multiple of 7 that can be subtracted* from it. Your goal is to *reduce* it to one of the numbers on the modulo 7 clock, namely 1, 2, 3, 4, 5, 6 and 0.

If you subtract a multiple of 7 that is *too small*, and the result is *still more than* 6, you will simply *subtract a multiple of 7 again*. There is no harm.

For example, in reducing 27, if you subtract 27 - 14, the result is 13 which still needs reducing. So you subtract again and $13 - 7 \equiv 6$ (mod 7). It's all good.

Example Reduce 47 in mod 7 with *repeated* subtraction of 7 *until the result is in modulo 7.*

$47 - 7 = 40 \rightarrow 40 - 7 = 33 \rightarrow 33 - 7 = 26 \rightarrow 26 - 7 = 19$

$19 - 7 = 12 \rightarrow 12 \equiv 5$ (mod 7).

47 slowly reduces to 5 in modulo 7. $47 \equiv 5$ (mod 7).

This can VERY time consuming.

If you *knew* that you should *subtract 42 from 47*, the problem is done quickly.

Find the equivalent *non-negative* values in mod 7 for the following.

The term *non-negative includes 0 and* means 0, 1, 2, 3, 4, 5 and 6 (in modulo 7).

Answers are below. NO PEEKING.

1) $16 \equiv$
2) $54 \equiv$
3) $63 \equiv$
4) $41 \equiv$
5) $99 \equiv$
6) $7 \equiv$
7) $3 \equiv$
8) $29 \equiv$
9) $36 \equiv$
10) $80 \equiv$
11) $19 \equiv$
12) $100 \equiv$
13) $64 \equiv$
14) $30 \equiv$
15) $32 \equiv$
16) $55 \equiv$
17) $40 \equiv$
18) $20 \equiv$
19) $69 \equiv$
20) $44 \equiv$

Answers:

1)	16 - 14 ≡ 2 (mod 7)	16 ≡ 2 (mod 7)
2)	54 - 49 ≡ 5 (mod 7)	54 ≡ 5 (mod 7)
3)	63 - 63 ≡ 0 (mod 7)	63 ≡ 0 (mod 7)
4)	41 - 35 ≡ 6 (mod 7)	41 ≡ 6 (mod 7)
5)	99 - 98 ≡ 1 (mod 7)	99 ≡ 1 (mod 7)
6)	7 ≡ 0 (mod 7)	7 ≡ 0 (mod 7)
7)	3 ≡ 3 (mod 7)	3 ≡ 3 (mod 7)
8)	29 - 28 ≡ 1 (mod 7)	29 ≡ 1 (mod 7)
9)	36 - 35 ≡ 1 (mod 7)	36 ≡ 1 (mod 7)
10)	80 - 77 ≡ 3 (mod 7)	80 ≡ 3 (mod 7)
11)	19 - 14 ≡ 5 (mod 7)	19 ≡ 4 (mod 7)
12)	100 - 98 ≡ 2 (mod 7)	100 ≡ 2 (mod 7)
13)	64 - 63 ≡ 1 (mod 7)	64 ≡ 1 (mod 7)
14)	30 - 28 ≡ 2 (mod 7)	30 ≡ 2 (mod 7)
15)	32 - 28 ≡ 4 (mod 7)	32 ≡ 4 (mod 7)
16)	55 - 49 ≡ 6 (mod 7)	55 ≡ 6 (mod 7)
17)	40 - 35 ≡ 5 (mod 7)	40 ≡ 5 (mod 7)
18)	20 - 14 ≡ 6 (mod 7)	20 ≡ 6 (mod 7)
19)	69 - 63 ≡ 6 (mod 7)	69 ≡ 6 (mod 7)
20)	44 - 42 ≡ 2 (mod 7)	44 ≡ 2 (mod 7)

It will be useful to *observe* that the <u>difference</u> between *two equivalent values in modulo 7* is *a multiple of 7*. Look at the first five answers.

16 ≡ 2 (mod 7) because 16 – 2 = 14 is a multiple of 7.
(and 14 is the multiple of 7 that was used to reduce it).

54 ≡ 5 (mod 7) because 54 – 5 = 49 is a multiple of 7.
(and 49 is the multiple of 7 that was used to reduce it).

63 ≡ 0 (mod 7) because 63 is a multiple of 7.

41 ≡ 6 (mod 7) because 41 – 6 = 35 is a multiple of 7.
(and 35 is the multiple of 7 that was used to reduce it).

99 ≡ 1 (mod 7) because 99 – 1 = 98 is a multiple of 7.
(and 98 is the multiple of 7 that was used to reduce it).

You can use subtraction to check if two non-negative numbers are *actually equivalent* in modulo 7.

Verify or challenge (and correct) the following:

- $50 \equiv 2$ (mod 7) can be *challenged and disproved.*
  The *difference* of 50 and 2 is 48.
  50 – 2 = 48 and we know 48 is *not* a multiple of 7.

- $101 \equiv 3$ (mod 7) can be *verified.*
  101 – 3 = 98 and we know 98 *is* a multiple of 7.

Verify or challenge these: Try not to look at the answers.

$43 \equiv 5$ (mod 7) $32 \equiv 4$ (mod 7) $83 \equiv 6$ (mod 7) $101 \equiv 2$ (mod 7)

Answers

- 43 - 5 = 38 *which is not a multiple of 7,* so $43 \equiv 5$ (mod 7) is *disproved.*
- 32 - 4 = 28 is a multiple of 7, so $32 \equiv 4$ (mod 7) is *verified.*
- 83 - 6 = 77 is a multiple of 7, so $83 \equiv 6$ (mod 7) is *verified.*
- 101 - 2 = 99, which is not a multiple of 7, $101 \equiv 2$ (mod 7) is *disproved.*

Negative Numbers and Negative Equivalents in Modulo 7

Sometimes, negative numbers come in handy. When we move 1 "hour" *counterclockwise* from the 0 position, we reach 6. If you think about the number line, when you move left from 0 what is the first number you reach? If you said -1, you are correct! The counterpart of moving *left on the number line* is moving *counterclockwise* on the modulo 7 clock.
$6 \equiv -1 \pmod 7$.

This is critical: The difference between the clock and the number line is that on the mod 7 clock, there are *always two answers to a problem; one negative and the other positive* (except for 0, of course).
At any of the six positive points on the modulo 7 clock, *two values can be assigned.* One is positive and the other is negative.

All positive numbers in modulo 7 (and in all modulo systems) have a negative equivalent. And all negative values have a positive equivalent. They are permanent partners. 6 and -1 are such partners and they are *equivalent in mod 7*. We state that fact by writing $6 \equiv -1 \pmod 7$.
Note: To verify this, *the sum of their absolute values* must be a multiple of 7. This is critical to retain. $6 + 1 = 7$ *verifies that $6 \equiv -1 \pmod 7$*.

IF THE SIGNS IN AN EQUIVALENCY ARE THE *SAME*, CHECK IT BY *SUBTRACTING* THE ABSOLUTE VALUES OF THE TWO NUMBERS. IF THE RESULT IS A MULTIPLE OF SEVEN, THEN THE EQUIVALENCY IS VERIFIED.

IF THE SIGNS IN AN EQUIVALENCY ARE _DIFFERENT_, CHECK IT BY _ADDING_ THE ABSOLUTE VALUES OF THE TWO NUMBERS. IF THE RESULT IS A MULTIPLE OF SEVEN, THEN THE EQUIVALENCY IS VERIFIED.

-11 ≡ 3 (mod 7) is an equivalency with _different signs._

This equivalency can only be verified by _adding_ the absolute values of -11 and 3:

11 + 3 = 14. Since 14 is _a multiple of 7_, this _verifies_ -11 ≡ 3 (mod 7).

What is the _negative equivalent of 4_ in mod 7? Complete this thought: 4 + ? = 7. _Immediately_ the answer is 3. Thus the _negative equivalent of 4 in modulo 7 is_ -3. Verifying 4 ≡ -3 (mod 7), _add_ their absolute values:

4 + 3 = 7, thus 4 ≡ -3 (mod 7) is verified.

For example, to find the _negative equivalent of 2_, complete this thought: 2 + ? = 7. How long did it take you to come up with 5? Not long at all, I hope.

Then -5 ≡ 2 (mod 7). _Adding_ 5 + 2 = 7 confirms that.

Only use this method of checking when _the signs of the numbers are different._

Verify or challenge each equivalency: The answers are just below. Don't look.

4 ≡ -3 (mod 7)   1 ≡ -5 (mod 7)   17 ≡ -4 (mod 7)   31 ≡ -5 (mod 7)   45 ≡ -4 (mod 7)

Answers:

- 4 ≡ -3 (mod 7) is verified: 4 + 3 = 7, which is a multiple of 7.
- 1 ≡ -5 (mod 7) is challenged: 1 + 5 = 6, which is _not_ a multiple of 7.
- 17 ≡ -4 (mod 7) is verified: 17 + 4 = 21, which is a multiple of 7.
  This may be easier if you reduced 17 _first_ to 3.
  3 ≡ -4 (mod 7) is verified _immediately_. 3 + 4 = 7.
- 31 ≡ -5 (mod 7): 31 + 5 = 36 which is _not a multiple of 7. Disproved._
  If you reduced 31 first, 3 ≡ -5 (mod 7): 3 + 5 = 8. 8 is _not a multiple_ of 7.
- 45 ≡ -4 (mod 7): 45 + 4 = 49, a multiple of 7. _Verified._
  If you reduced 45 first, 3 ≡ -4 (mod 7) is verified _immediately._

Find the negative equivalent for each value. The answers are below. No peeking. You will see that *the negative sign is interchangeable.*

The negative equivalent of 2 is   The negative equivalent of 1 is
The negative equivalent of 3 is   The negative equivalent of 5 is
The negative equivalent of 7 is   The negative equivalent of 6 is
The negative equivalent of 4 is   The negative equivalent of 0 is

Answers:

The negative equivalent of 2 is -5   The negative equivalent of 1 is -6
The negative equivalent of 3 is -4   The negative equivalent of 5 is -2
The negative equivalent of 7 is 0   The negative equivalent of 6 is -1
The negative equivalent of 4 is -3   The negative equivalent of 0 is 0

Note: $2 \equiv -5$ *(mod 7) and* $5 \equiv -2$ *(mod 7)* which illustrates that the negative sign is interchangeable between a positive number in modulo 7 and its negative equivalent.

Is $57 \equiv 3$ (mod 7)? *Neither is negative,* so find the *difference.*
57 - 3 = 54 but is 54 a multiple of 7? No! Then $57 \equiv 3$ (mod 7) is *disproved.* You can check equivalencies like these *all the time.*

On the *modulo 7* clock, here are six equivalencies:
$2 \equiv -5$ (mod 7) $1 \equiv -6$ (mod 7) $3 \equiv -4$ (mod 7) $5 \equiv -2$ (mod 7) $6 \equiv -1$ (mod 7)
$4 \equiv -3$ (mod 7)

You don't need to remember all of these to remember negative equivalents: All you need is *three pairs*:
$1 \equiv -6$ (mod 7) $2 \equiv -5$ (mod 7) $3 \equiv -4$ (mod 7)
This is because in an equivalency, a single negative sign is interchangeable.

Verify or challenge the following. Pay attention to *the number of negative* signs! Answers follow. Don't look.
1) $17 \equiv 3$ (mod 7)   2) $-5 \equiv -16$ (mod 7)   3) $-6 \equiv 1$ (mod 7)

4) $-19 \equiv -2$ (mod 7)   5) $-12 \equiv 5$ (mod 7)

Answers

(1)   True because 17 - 3 is 14, which is not *a multiple of* 7.

(2)   False because 16 - 5 = 11, *not a multiple of* 7. (The signs are the same!)

(3)   True because 6 + 1 = 7, which is a multiple of 7.

(4)   False because 19 - 2 = 17, which is *not a multiple of* 7.

(5)   False because 12 + 5 = 17, which is *not a multiple of* 7.

Reducing a Negative Number in Modulo 7

Negative numbers can be turned into positive numbers in modulo 7. You can also convert 6, 5 and 4 to -1, -2 and -3 whenever it suits you. Also, like a balloon, we can "blow up" a number by adding a suitable multiple of 7. But you have to keep that multiple of 7 as *small* as possible to reduce a *negative number*. Usually you want to *maximize* the multiple of 7. That is when you are reducing a *positive number*. To reduce -32, find the *smallest multiple of 7 that can be <u>added</u> to -32* in order to get a *positive* result.

Is it 7? 14? 21? Stop at the *first (and smallest) multiple of 7 that is more than 32*. In this way, any negative number in modulo 7 can be converted to an *equivalent positive* number.

If you *add -32 + 35 = 3*, you have found the *positive* equivalent of -32: -32 ≡ 3 (mod 7).

Reduce -12 in mod 7
We need the *smallest multiple of 7* that we can add to -12 that will have a *positive* result.
Start *counting by 7's* until you reach a multiple of 7 that is *greater than 12*. It shouldn't take you long to find 14. Then add: -12 + 14 = 2. You are done.

Maybe you can find another way to do these problems.
1) Reduce - 43 in mod 7   2) Reduce -51 in mod 7   3) Reduce –18 in mod 7

Answers follow.

Answers and solutions: The multiple of 7 required is shown.

1) Counting by 7's: 7, 14, 21, 28, 35, 42, and <u>49</u> STOP.
    49 is the first multiple of 7 that is greater than 43. Adding -43 + 49 = 6.
    You are done.
    -43 ≡ 6 (mod 7). Verify: 43 + 6 = 49, a multiple of 7.

2) Counting by 7's: 7, 14, 21, 28, 35, 42, 49, and <u>56</u> STOP.
    56 is the first multiple of 7 that is greater than 51. Adding -51 + 56 = 5.
    You are done.
    -51 ≡ 5 (mod 7). Verify: 51 + 5 = 56, a multiple of 7.

3) Counting by 7's: 7, 14, and <u>21</u> STOP.
    21 is the first multiple of 7 that is greater than 18.
    Adding -18 + 21 = 3. You are done.
    -18 ≡ 3. Verify: 18 + 3 = 21, a multiple of 7.

Of course, you can start counting with any multiple of 7 you like. As long as it is less than the number you are trying to reduce. Know the multiples of 7.

Negative numbers can be very useful at times.
To perform an operation like 6 + 5, it *can be* useful to <u>see</u> (and *think* of) 6 as -1.
6 + 5 can be re-written as -1 + 5 ≡ 4 (mod 7). The -1 can *replace* 6 in modulo 7.
For 4 - 5, -5 can be replaced by 2 (or +2) because -5 ≡ 2 (mod 7).
Thus, 4 – 5 can be re-written 4 + 2 ≡ 6 (mod 7).

If you are faced with the problem 5 - 6 (mod 7), you can "blow 5 up" to make it *appear* larger by adding a multiple of 7. That is an *illusion* in modulo 7 as we remarked before. The balloon you actually blow up is the same balloon as it was deflated. It just contains air now. So 5 becomes 5 + 7 = 12.
Then 5 - 6 becomes 12 - 6 = 6.

Alternately, you can find the positive equivalent of -6 which is 1 or + 1.
Thus, 5 – 6 becomes 5 + 1 and the answer is *still* 6 (mod 7).

Only in modulo arithmetic can you can *redesign problems* to suit yourself!

Example: 4 - 16 (mod 7)
4 can be "blown up" by adding 14 to 4. Thus, 4 + 14 = 18.
Thus, 4 − 16 becomes 18 -16 ≡ 2 (mod 7). We are done.
Alternately, -16 can be made *positive* by adding a *suitable multiple* of 7.
The multiple 21 does the job and -16 + 21 ≡ 5 (mod 7).
Thus, 4 - 16 becomes 4 + 5 and 4 + 5 ≡ 9 ≡ 2 (mod 7).
You get the same answer, 2.

Or, if you are good with signed numbers, you can do *conventional signed number arithmetic*: 4 - 16 = -12 and *then reduce -12 in* modulo 7.
You can "blow up" -12 by *adding 14, a multiple of 7*. -12 + 14 = +2 or 2. We are done.
There will be *many* ways to do the problems in this book. Experiment.

Example: 8 − 19
8 can be "blown up" to a number that is *greater* than 19. Add a multiple of 7.
14 does the job and 8 + 14 = 22. The problem is now 22 − 19 = 3. We are done.
Alternately, -19 can be made *positive* by adding a suitable multiple of 7.
The multiple 21 does the job and -19 + 21 = 2 or +2.
Then 8 − 19 becomes 8 + 2 ≡ 10 ≡ 3 (mod 7).

Or you can do *conventional signed number arithmetic*: 8 -19 = -11 then reduce *-11*.
-11 can be "blown up" by adding 14. Adding 14: -11 + 14 = 3.
We get the same answer, naturally. Modulo 7 is *very* flexible.

Once you are versatile with negative numbers you will find yourself replacing 6 with -1 or 5 with -2 or 4 with -3 or 7 with 0 or 8 with 1 etc. when it is convenient. You wouldn't usually replace -4 with 3 but you can if it suits you. Replacing 6, 5 and 4 with negative numbers causes the other positive numbers in the problem to decrease in value. The more mental computation you do the faster the system works. *Think of yourself as a number processor.* There is more on replacing numbers in problems in the next section, Reducing in Modulo 7 As You Work.

Reducing in modulo 7 as you work

When working in modulo 7, look carefully *for combinations that are equivalent to 0* (like 4 + 3 or 5 + 2 or 6 + 1) or any other *recognizable* values in modulo 7 hidden in the problem (in plain sight).

If you recognize that $12 + 2 \equiv 0 \pmod 7$, then 12 + 3 + 2 *immediately* can become 3 in modulo 7 after deleting 12 and +2 which combined, is equivalent to zero. *I call this "scratching zeroes".*

The problem gets whittled down.

In 12 + 3 + 2, someone else might see 12 + 3 as 1 in modulo 7. This is because $12 + 3 = 15 \equiv 1 \pmod 7$. *Then* 1̲2̲ ̲+̲ ̲3̲ *+ 2 morphs to* 1̲ *+ 2 = 3.* Here, reducing is unnecessary.

In another case, 2 – 6 can *immediately* become *2 + 1 = 3* in modulo 7. This is because $-6 \equiv 1 \pmod 7$. Keep watch. The numbers are your friends. And they are our pet chameleons. Negative numbers can be changed to equivalent positive numbers. Positive numbers can be converted to equivalent negative numbers. How cool is that?

Try a few exercises. You should explore other ways to do them. Modulo arithmetic is like silly putty. You can mold the problem to your liking. Solutions shown are *not unique*. Try other approaches. Be original.

6 + 23 + 4 (mod 7). *How fast* can you get the result? Reducing mentally is a powerful tool.

Notice that 23 can be reduced immediately to 2 [since $23 – 21 \equiv 2 \pmod 7$]. So the problem is now *6 + 2 + 4 (mod 7).*

I also notice that 6 is equivalent to -1 (mod 7). Using that substitution has the effect of making another number in the problem smaller.

Now the problem is *-1 + 2 + 4* (mod 7).

The numbers have been whittled down.
Using conventional signed number arithmetic, the result is 5.

Try other ways to re-shape the problem to make it easier for you. If you can do these changes, much of the arithmetic in modulo 7 can be mental.

Find the following sums and reduce in mod 7. Use shortcuts.

1) $5 + 7 + 2 \equiv$ (mod 7)   2) $4 + 43 + 6 \equiv$ (mod 7)   3) $16 + 5 + 5 \equiv$ (mod 7)

4) $2 + 30 + 6 \equiv$ (mod 7)   5) $12 + 22 + 35 \equiv$ (mod 7)   6) $40 + 19 + 6 \equiv$ (mod 7)

Answers follow but don't look at them yet.

There are many ways to do these problems. I am using a fictional "alert student", Alex.
Another technique of reducing is *selectively reducing the digits* of a number.
Of course, the digits you reduce must be 7, 8 or 9.
For example, in reducing 19, the 9 can be reduced to 2 changing 19 to $12 \equiv 5 \ (mod \ 7)$.
Or in reducing 95, you can instead reduce $25 \equiv 4 \ (mod \ 7)$.
There will be more on reducing larger numbers in a later chapter.

Answers: Alex will be an alert student doing these problems. No two people are likely to solve these the exact same way. Be original. Experiment.

1) 0   Alex would *scratch* 7 because *7 is equivalent to 0 (mod 7)*.
        Alex would ALSO *scratch* 5 + 2 because *5 + 2 ≡ 0 (mod 7)*.
        ALL numbers are gone. Then *5 + 2 = 7 ≡ 0 (mod 7)*.

2) 4   Alex sees that 43 ≡ 1 (mod 7). Also, 6 ≡ -1 (mod 7).
        Then the problem is 4 + 1 - 1 ≡ 4 (mod 7).

3) 5   Alex sees that 16 ≡ 2 (mod 7).
        The student scratches 2 + 5 because it is equivalent to 0 (mod 7).
        The result is 5 still standing.

4) 3   Alex sees that 30 ≡ 2 (mod 7). Also, 6 ≡ -1 (mod 7).
        Now the problem is: 2 + 2 - 1 ≡ 3 (mod 7).

5) 6   Alex sees that 12 ≡ 5 (mod 7) and that 22 ≡ 1 (mod 7).
        The student *scratches* 35 because 35 ≡ 0 (mod 7).
        Now the problem is: 5 + 1≡ 6 (mod 7).

6) 2   Alex sees that 40 ≡ 5 (mod 7) and 19 ≡ 5 (mod 7).
        The student *observes* that 6 ≡ -1 (mod 7).
        Now the problem is: 5 + 5 -1 = 9 ≡ 2 (mod 7).

See how many ways you can manipulate the numbers in these problems to your advantage.

Solving equations in modulo 7

We have already seen a few equations in the review of algebra.
$x + 2 \equiv 6 \pmod 7$ can be solved by transposing $+2$ to the right side of the $\equiv$ sign.
Thus $x \equiv 6 - 2 \equiv 4 \pmod 7$.

You will be relieved to know that $x + 6 \equiv 2 \pmod 7$ can be done in the same exact way.
$x \equiv 2 - 6 \equiv -4 \equiv 3 \pmod 7$. The solution to this equation is 3.
The use of negative equivalents will be invaluable.

Example 1. Solve $x + 4 \equiv 2 \pmod 7$:
The arrow $\rightarrow$ separates steps in the evolution of the equation as it changes.

$x + 4 \equiv 2 \pmod 7 \rightarrow x \equiv 2 - 4 \rightarrow x \equiv -2 \pmod 7 \rightarrow x \equiv 5 \pmod 7$
The solution is 5.

Example 2. Solve: $x + 2 \equiv 4 \pmod 7$.
$x + 2 \equiv 4 \pmod 7 \rightarrow x \equiv 4 - 2 \rightarrow x \equiv 2 \pmod 7$
The solution is 2.

Example 3. Solve: $x + 3 \equiv 1 \pmod 7$
$x + 3 \equiv 1 \pmod 7 \rightarrow x \equiv 1 - 3 \equiv -2 \pmod 7 \rightarrow x \equiv 5 \pmod 7$
The solution is 5.

Example 4. Solve: $x + 3 \equiv 0 \pmod 7 \rightarrow x \equiv 0 - 3 \rightarrow x \equiv -3 \pmod 7 \rightarrow x \equiv 4 \pmod 7$

Example 5. Solve $x + 5 + 3 \equiv 1 \pmod 7$
This requires a little more thought because it involves more numbers.

Combine 5 + 3 on the *left side* as we did before in the review:

Solve $x + 8 \equiv 1 \pmod 7$.

Here is the evolution of the equation as it changes form.

$x + 5 + 3 \equiv 1 \pmod 7 \rightarrow x + 8 \equiv 1 \pmod 7 \rightarrow x + 1 \equiv 1 \rightarrow x \equiv 1 - 1 \pmod 7$

Finally, $x \equiv 0 \pmod 7$

The solution is 0.

Solve the following:

1) $x + 4 + 3 \equiv 2 \pmod 7$

2) $x + 6 + 3 \equiv 4 \pmod 7$

3) $x + 3 + 0 \equiv 5 \pmod 7$

4) $x + 6 + 5 \equiv 0 \pmod 7$

5) $x + 0 + 2 \equiv 6 \pmod 7$

6) $x + 5 + 0 \equiv 4 \pmod 7$

7) $x + 1 + 4 \equiv 5 \pmod 7$

8) $x + 3 + 6 \equiv 0 \pmod 7$

9) $x + 5 + 5 \equiv 5 \pmod 7$

10) $x + 6 + 1 \equiv 3 \pmod 7$

Solutions: *Experiment with your own solutions. The last equation can be compacted.*

1) $x+4+3\equiv2\,(\text{mod}\,7)\rightarrow x+7\equiv2\,(\text{mod}\,7)\rightarrow x+0\equiv2\,(\text{mod}\,7)\rightarrow x\equiv2\,(\text{mod}\,7)$

2) $x+6+3\equiv4\,(\text{mod}\,7)\rightarrow x+9\equiv4\,(\text{mod}\,7)\rightarrow x+2\equiv4\,(\text{mod}\,7)\rightarrow x\equiv2\,(\text{mod}\,7)$

3) $x+3+0\equiv5\,(\text{mod}\,7)\rightarrow x+3\equiv5\,(\text{mod}\,7)\rightarrow x\equiv5-3\equiv2\,(\text{mod}\,7)$

4) $x+6+5\equiv0\,(\text{mod}\,7)\rightarrow x+11\equiv0\,(\text{mod}\,7)\rightarrow x+4\equiv0\,(\text{mod}\,7)$
   $x\equiv-4\equiv3\,(\text{mod}\,7).$

5) $x+0+2\equiv6\,(\text{mod}\,7)\rightarrow x+2\equiv6\,(\text{mod}\,7)\rightarrow x\equiv6-2\equiv4\,(\text{mod}\,7)$

6) $x+5+0\equiv4\,(\text{mod}\,7)\rightarrow x+5\equiv4\,(\text{mod}\,7)\rightarrow x\equiv4-5\equiv-1\equiv6\,(\text{mod}\,7)$

7) $x+1+4\equiv5\,(\text{mod}\,7)\rightarrow x+5\equiv5\,(\text{mod}\,7)\rightarrow x\equiv5-5\equiv0\,(\text{mod}\,7).$

8) $x+3+6\equiv0\,(\text{mod}\,7)\rightarrow x+9\equiv0\,(\text{mod}\,7)\rightarrow x\equiv0-9\,(\text{mod}\,7)\rightarrow x\equiv-9\,(\text{mod}\,7)$
   Now find a suitable multiple of 7 to add to -9 that will make the result positive.
   -9 + 14 = 5. The solution is 5. Try other ways to solve this.

9) $x+5+5\equiv5\,(\text{mod}\,7)\rightarrow x+10\equiv5\,(\text{mod}\,7)\rightarrow x\equiv5-10\,(\text{mod}\,7)$
   $x\equiv-5\equiv2\,(\text{mod}\,7)$

10) $x+6+1\equiv3\,(\text{mod}\,7)\rightarrow x+7\equiv3\,(\text{mod}\,7)\rightarrow x\equiv3\,(\text{mod}\,7)$
    (7 vanished because it is equivalent to 0.)

The Basics of the Gregorian calendar

Years that are divisible by 4 are called Leap Years. There *is an exception*.

Years ending with two zeroes, like 1800 or 1900, are *not* ordinarily leap years even though they *seem* to qualify, as they are divisible by 4. Nevertheless, they are *not always leap years* in the Gregorian calendar.

Centennial years are, however, *are* leap years *because they are divisible by 400*. Such years all *end a century* that precedes a *leap century*. The term "turn of the century" will be used loosely when referring to such years. Forgive the author in advance, if a year such as 1800 is referred as the start of the century or the *turn of the century*.

The year 2000 was a leap year which *ended* the 20th century and ushered in the 21st century, a *leap century*. The last such centennial year was 1600 (*so the 17th century was a leap century*). These facts are *not essential* for the tasks to come. The next centennial year will be 2400 so *the 25th century will be a leap century*. Don't wait up for it.

In moving from year to year, the *day of the week* on a which a date falls normally skips to the *next* day of the week in the *following year*, with this *exception*; if the following year is a *leap year, two days are skipped* (unless the date falls in *January or February*; more on that later).

For example, March 1, 1880 falls on a Monday. Trust me.
March 1, 1881 subsequently falls on a *Tuesday*.
This is the *normal effect* of moving from *one year to the next non-leap year*.
The day of the week moves up to the *next* day of the week.

Now let's take a look at March 1, 1879, which falls on a *Saturday*.
On what day of the week should it fall in 1880? It was not Sunday. We saw
that March 1, 1880 falls on a Monday. So why, in 1880, did March 1, 1880
skip an extra day to Monday?
The reason is *1880 was a leap year.*

Any dates *after February 29 are affected by leap years.*
Dates in January and February are not affected by this phenomenon.

In that light, examine the dates *February 12, 1879 and February 12, 1880.*
Dates in January or February behave differently than the other months.
Moving into a leap year, there is no "leap year effect" of two days skipped.
February 12, 1879 falls on a *Wednesday.*
Moving into 1880, a leap year, February 12, 1880 falls on a *Thursday.*
There was only the normal skip of one day. There is no leap year effect.
(Later on you will be able to find these days of the week yourself!)
Any date preceding March is unaffected by the phenomenon of a leap year. This is
extremely important to know.

Another example: January 18, 1879 falls on a Saturday.
January 18, 1880 falls on a Sunday, *one day* of the week after Saturday.
Only one day was skipped even though we went into a leap year! This is because
the leap had not happened yet, that is, we hadn't reached March 1, 1880 yet.

An exercise: *Given* that December 3, 1975 falls on a Wednesday, find the day
of the week on which the following dates fall.

1) December 3, 1977

2) December 3, 1979

3) December 3, 1980

4) December 3, 1978

Answers:

1) From 1975, add two days to reach 1976 because 1976 is a *leap year*.
    Skipping one *more* day brings us to 1977. We have advanced three days.
    Three days after Wednesday is Saturday. Using codes, the code for
    Wednesday is 4. Then $4 + 3 = 7 \equiv 0$ (mod 7).
    0 is the code for Saturday. December 3, 1977 falls on a Saturday.

2) We can use 1977 as a jump off year to reach 1979. From 1977 add one
    day to reach 1978. Add one *more* day to reach 1979. We did not pass a
    leap year. There is a total of two days skipped. Two days after Saturday
    is Monday. December 3, 1979 falls on a Monday.
    Using codes, the code for Saturday is 0. $0 + 2 = 2$, which is the code for Monday.

3) We can use 1979 as a jump off year to reach 1980. From 1979 there are
    two days skipped to reach 1980 because 1980 is a *leap year*. Two days
    after Monday is Wednesday.
    Using codes, the code for Monday is 2. $2 + 2 = 4$, the code for Wednesday.
    December 3, 1980 falls on a Wednesday.

4) To reach 1978, we will use the date in 1977 as a jump off point.
    Add one day to reach 1978. One day after Saturday is Sunday.
    December 3, 1978 falls on a Sunday.

In a short time, you will be to calculate the day of the week for any date without
resorting to counting like this.

The Main Number for Years is 5

In any normal century, that is, not a *leap century*, there are 100 *leap days* that normally occur from one non-leap year to the next non-leap year. Also, any *normal* century has *24 leap years*. Note: there *are only 24 leap years, not 25*, because a year ending in 00 is *usually* not a leap year. *Leap centuries, however,* have *100* leap days and <u>*25*</u> *leap years.*

1900 ushered in the 20ᵗʰ century but 1900 *was not a leap year*. 2000 is a leap year and thus it is a centennial year. The 21ˢᵗ century is a leap century.

We add the number of 100 normal leaps and the number of 24 *extra leaps* using modulo 7.

100 + 24 (mod 7) is the problem at hand. Do it now. Your result will be the same.

The long way is to add 100 + 24 = 124 *then reduce 124* in modulo 7.
The *largest multiple of 7 less than 124 is 119*. Then subtract:
$124 - 119 \equiv 5 \pmod 7$.
Thus the result is 5.
For leap centuries, that number is 6 because there are 25 leap years.

Or we can start by *reducing both 100 and 24 in modulo 7*.
To reduce 100 in modulo 7, we subtract 98 from 100.
100 normal leaps reduces to 2 (mod 7) and 24 *extra leaps* reduces to 3 *(mod 7)*.
Add the two results: $2 + 3 \equiv 5 \pmod 7$.
The result is 5 leaps for each normal century.
We have the same result, 5.

This number 5 is critical for all centuries.
5 is the number of weekdays skipped for any date moving from one normal century to the next normal century. Know that in moving a date from a _normal century_ to a _leap_ century, the number of weekdays skipped is 6.
This is VERY IMPORTANT. Make sure you retain this.

We will now _explain and apply_ it. A date in the 19th, 20th and 21st century will clarify this.
April 15, 1865 falls on a Saturday whose code is 0. Lincoln was assassinated.
In the 20th century, April 15, 1965 falls on a Thursday whose code is 5, _an advance of 5 days_ from the Saturday in 1865.
This is what _normally_ happens to a date from one normal century to the next normal century.

Moving to the next century, the 21st century is a _leap century._
April 15, 2065 falls on a Wednesday, whose code is 4, _an advance of 6 days from Thursday in 1965._ We went from Saturday in 1865 to Thursday in 1965 and then to Wednesday in 2065. We made a leap of 5 days, and _then_ a leap of 6 days. A total of 11 days were skipped from the Saturday in 1865 to Wednesday in 2065. 11 is equivalent to 4 in modulo 7 and 4 days after Saturday is Wednesday. An extra day was skipped because the 21st century is a leap century.

Another demonstration:
April 15, 1865 falls a Saturday whose code is 0.
April 15, 1965 skips five weekdays to Thursday, whose code is 5. $0 + 5 = 5$.
Thus, April 15, 1965 falls on a Thursday.

April 15, 2065, skips six weekdays from Thursday. It jumps from Thursday (in 1965) to Wednesday. $5 + 6 \equiv 11 \equiv 4 \pmod 7$.
Thus, April 15, 2065 falls on a Wednesday.
Also, instead of adding 6, we can subtract 1 from the code of Thursday.
Thus $5 - 1 = 4$, the code for Wednesday.

Exercise: Given that November 9, 1782 falls on a Saturday, on what day of the week does November 9, 2082 fall?

Moving from the 18th century to the 19th century, we *skip five* days.
Five days after Saturday is Thursday. November 9, 1882 falls on a Thursday.
Moving from the 19th century to the 20th century, another five days are skipped.
Five days after *Thursday* is Tuesday. November 9, 1982 falls on a Tuesday.

Finally, moving from the 20th century to the 21st century, *six* days are skipped because the 21st century is a leap century. *Six* days after *Tuesday* is Monday. November 9, 2082 falls on a Monday.

You can also use codes to solve this problem.
After calculating the *total* number of skipped days, add.
We skipped $5 + 5 + 6 = 16$ days.
The code for Saturday (November 9, 1782) is 0.
Thus, adding $0 + 16 \equiv 2 \pmod 7$ indicates 2, the code of Monday, is the answer.

The Mathematics of the Calendar

Someone devised a system that can (among other things) tell the day of the week of *ANY DATE* (as long as the date is in the Gregorian calendar).

Recall the days of the week have codes in modulo 7:

Sunday has the code 1   Monday has the code 2   Tuesday has the code 3
Wednesday has the code 4   Thursday has the code 5   Friday has the code 6
Saturday has the code 0 (*not* 7)

The reader is urged to memorize this sequence. You just have to know that the code for Sunday is 1 and that the code after 6 is 0.
Learn to think "*the day after 6 is 0*".

This system was developed in the 19th century.
The creator of this system knew the following facts:
1. Every century *must have a code* from 0 to 6. Sound familiar? We *are* in modulo 7.
2. Every *year* must have a code from 0 to 6.
3. Every *month* must have a *code* from 0 to 6.

The creator decided to start with the code 0 for the 19th century.
Thus, 0 is the code for 1800.

Remember, *leap centuries only occur every 400 years*. The next leap century will be the 25th century which is ushered in with the centennial year 2400.

Century Codes

Now we will put the number 5 to work. It will find the century codes for us.

The code for the 19th century is 0 *(by default)*. The century codes summarized: We use the fact that dates advance five days of the week from a century to the next century, *(as long as the next century is not a leap century!)*. *Thus, the* code for the 20th century is $0 + 5 \equiv 5 \ (mod \ 7)$.

The 21st century is *a leap century*.

Now we must add 6 to 5 or *subtract 1 from 5*.

The code for the 21st century is thus $5 + 6 \equiv 11 \equiv 4 \ (mod \ 7)$ or $5 - 1 \equiv 4 \ (mod \ 7)$. It might interest some readers to know that *all leap centuries* have the code of 4.

The code for the 22nd century resumes the five days advance; $4 + 5 \equiv 9 \equiv 2 \ (mod \ 7)$.

These codes, 0, 5, 4 and 2, are the **only century codes**. Each one repeats every 400 years or four centuries.

The code for the 19th century is 0.

In four centuries), the code for the 23rd century will also be 0.

It's easy to get the codes of non-leap centuries by *adding 5 going forward* or *subtracting 5 going back in time*. You also have the option to subtract 2 going *forward* and add 2 going *backward*. This is possible because $5 \equiv -2 \ (mod \ 7)$.

Moving *forward*, this pattern becomes clearer. The code for the 16th century is 5 (at least for part of it).

Since the 17th century is a *leap century*, we *add 6 to get its code*.

$5 + \mathbf{6} \equiv 4 \ (mod\ 7)$.

*Subtracting 1 has the same result. 5 - **1** = 4.*

Then the code for the 17th century is 4.

Moving on to the 18th century, we *add 5. Thus, 4 + **5** ≡ 2 (mod 7), so the code for the 18th century is 2. Subtracting 2 has the same result. 4 - **2** = 2.*

The code for the 18th century is 2. Continue to add 5 (or subtract 2) reach the codes of the 19th and 20th centuries which are 0 and 5, respectively. After the 20th century, we add 6 to find the code of the 21st century.

Here it is, all laid out.

The 16th century code is 5. The 17th century code is 4.

The 18th century code is 2.

The 19th century code is 0. The 20th century code is 5.

The 21st century code is 4.

The 22nd century code is 2. The 23rd century code is 0.

The 24th century code is 5.

The 25th century code is 4. The 26th century code is 2.

A cycle of 0, 5, 4, and 2 of century codes perpetually repeats itself.

After 2, the cycle repeats with 0 5 4 2 0 5 4 2 ad infinitum.

These are *the only century codes*.

This information on centuries is to satisfy your curiosity (if you have it) about the codes for centuries. You do not need it. It is enough that you know the code for the 20th century is 5 and the code for 21st century is 4. It is unlikely that you will be called upon to solve a problem that requires the code for the 19th or 18th century but their codes, 0 and 2, are easy to remember.

The *Year Codes: The messy part*

Before we begin, it is essential that you are able to *recognize* a *leap year* when you see one. All you need is to examine *the last two digits of the year*. If they form a number that is *a multiple of 4*, then it is a leap year. Otherwise, it is *not*. Your first task is to learn how to find the day of the week for *any* date in any year in any century. You *will* be able to do that after becoming fluent in modulo 7.

No one can be expected to know *all* of the year codes. Happily, there is a way to calculate them from scratch with knowledge of modulo 7 and basic addition. It is a process called the *year algorithm* and it will be explained later. To begin the process of finding the day of the week for a date, you *need the code of the century in which the year appears*. That isn't so bad since centuries aren't changing all the time. But years do. You should already know some century codes. In most cases the century will be the 20th or the 21st. You will probably only have to know the codes of those two centuries, 5 and 4, respectively.

Since 1800 has the code 0, it set the stage for subsequent years.
1801 has code 1, 1802 has code 2, 1803 has code 3 *but 1804 was a leap year.*

Leap years are *annoying* because they have <u>two codes</u>.
The first code + 1 ≡ the second code (mod 7).
Since the code for 1803 is 3, the *first* code for 1804 is 4.
The *second* code for 1804 is 4 + 1 = 5.
If the first code of a leap year were 6, then $6 + 1 \equiv 7 \equiv 0$ (mod 7) indicates *the second code* is 0.

Note: This is the *only* instance in which you <u>need</u> modulo 7 to find the second code for a leap year. In all other cases, it will be a simple elementary addition of 1, like 3 + 1 = 4.

Leap years are made up of two separate parts.

The first part is *January and February.*

In 1804, the code for the *first* part is 3.

The second part is March through December.

The code for *the second* part of 1804 is 5.

Throughout this book references will be made to *the first code and the second code* of a leap year. It is *not* obvious that 6 is 1 less than 0 in modulo 7. The modulo 7 clock, as we discussed before, is circular. If a race car driver on a circular track sees a car in his rear view mirror, it does not necessarily mean that he is leading in the race. In fact, the car "behind" him may have completed more laps and is therefore ahead. If it were a drag race on a straight course, then the car behind him would *definitely lose.*

1804 has the dual code (4, 5).

On Wednesday, February 29, 1804, the code for 1804 is 4, *the first code.*

On Thursday, March 1, 1804 the code for 1804 is 5 and the *leap* occurs.

2000 *is* a leap year. *It has two codes*; 3 and 4. The *second code* of 2000 sets the code for the 21st century. Calculating year codes will be in a later chapter.

Soon you will be able to *calculate* year codes and use the year codes.

Finding Year Codes Locally by Counting

By locally, we mean within about 10 years. If you know the code of a year, say 1997, and you want the code of a *subsequent* year, you can simply *count on your fingers*, 1 year at a time until you reach that year. It sounds simpler than it is. There are obstacles.

For the sake of simplicity, let's say a date in question occurs *after February*. This eliminates leap years as part of our problem. We will discuss leap years in depth, later.

Suppose you need the code for 2002.

You can "leap frog" from one year to the next year. We do so by *repeated addition* of 1. This needs to be done carefully. Every fourth year is a leap year and it's easy to pass one. Leap years are like mines.

When you are adding 1 and you reach a leap year, pause.

State two codes. State the *first code* and then state the *second code*.

Start with the code of 1997 which is 0. We Add 1.

Thus, $0 + 1 = 1$, which is the code for 1998.

Then we add 1 *again* to the get the code for 1999.

Thus, $1 + 1 = 2$, *which is the code for 1999.* So far, so good. No leap year *yet*.

Again, we add 1 to the code of 1999.

Thus, $2 + 1 = 3$, *which is the first code for 2000,* a leap year.

Pause. You have found that the *first code* of 2000 is 3.

Thus, the *second code* is $3 + 1 = 4$.

The dual codes for 2000 are *3 and 4*. Now you can resume counting from 4.

Thus, 4 + 1 = 5 which is the code for 2001. Then 5 + 1 = 6, the code for 2002. *We have found the code for 2002. It is 6.* We can *do this repeated addition* at most *three times* before we reach *another leap year.*

Another example: Start from the code of 2002 which is 6.

Add $6 + 1 \equiv 7 \equiv 0 \pmod{7}$.

Then the code for 2003 is 0. Add 1 to the code for 2003. Then 0 + 1 = 1.

We just entered 2004, a leap year. Pause.

The *first code* of 2004 is 1. The *second code* is 2. The *dual code* for 2004 is (1, 2).

Be careful. Below shows the progression of codes from 2000 to 2004.

2000 codes (3, 4) → 2001 code 5 → 2002 code 6 → 2003 code 0 → 2004 codes (1, 2)

I *always* use my fingers to do this. If anyone watches you do that, there will still be a mystery about *what* you are doing. *We can do this counting indefinitely.* If a year is too far ahead, and this method not feasible, there is a *year algorithm* that can be used to calculate the year code. (An algorithm is a step by step procedure to solve a problem.)

Warning: The result of the year algorithm for a leap year, will <u>always</u> be *the* <u>second</u> code. *You will have to subtract 1 to find the first code.*

Finding Year Codes
Using the Year Code Algorithm

This is a four step year algorithm: We will use non-leap years to start
We will use 1945 as an example. Bullets will emphasize each step.

We only use the *last two digits of the year*. 45 are the last two digits of 1945.

* 1. Divide 45 by 4, and *drop the remainder*.
     There will always be a remainder for a non-leap year!
     $45 \div 4 = 11$ Reduce 11.   *$11 \equiv 4 \ (mod \ 7)$*
* 2. *Reduce 45 in modulo* 7. $45 - 42 \equiv 3 \ (mod \ 7)$.
     So far, the two results are 4 and 3.
* 3. Acquire the code for the century in which 1945 lies.
     The *20th century has code 5*.
* 4. The results are 4, 3 and 5. Now add the three results *in modulo* 7.
     $4 + 3 + 5 \equiv 12 \equiv 5 \ (mod \ 7)$. Reducing is not always necessary.

Thus, we have found that the code for 1945 is 5.

Another example of a *non-leap year* with *abbreviated steps*: 1954

* $54 \div 4 = 13$ Reduce 13 in modulo 7. $13 \equiv 6 \ (mod \ 7)$
* $54 \equiv 5 \ (mod \ 7)$
* The code for the 20th century is 5.
* The sum of the results from the three steps is $6 + 5 + 5 \equiv 16 \equiv 2 \ (mod \ 7)$.

The code for 1954 is 2. There are many other ways to do this addition (in mod 7).

Using the *year code algorithm* for a *leap year*

An example of finding the <u>two</u> codes for *a leap year*:
Find the two codes for 1964

❖ $64 \div 4 = 16 \equiv 2 \pmod 7$ (**there is no remainder this time**)
❖ $64 \equiv 1 \pmod 7$
❖ The code for the 20th century is 5.
❖ The sum of the results of the three steps $2 + 1 + 5 \equiv 8 \equiv 1 \pmod 7$.
1 is the *second* code for 1964. Subtract 1- 1 to find the *first code*.

Thus, $1 - 1 = 0$, the first code for 1964. *Then the dual codes for 1964 are 0 and 1.*

This method is in contrast to the method of *counting* with your fingers whereby you find the *first code* and <u>then</u> the second code follows.

Find the two codes for 1980.

❖ $80 \div 4 = 20 \equiv 6 \pmod 7$
❖ $80 \equiv 3 \pmod 7$
❖ The code for the 20th century is 5.
❖ The sum of the results of the three steps $6 + 3 + 5 \equiv 14 \equiv 0 \pmod 7$.

0 is the *second* code for 1980. Subtract 0 - 1 to find the *first code*.
Thus, $0 - 1 \equiv -1 \equiv 6 \pmod 7$ *Then the dual codes for 1980 are 6 and 0.*

So the dual codes for 1980 are 6 and 0. (Also they are the codes for 2008).
If you *calculate* a code for a *leap year*, you must perform the *extra step of subtraction*.

Find the codes for the following years. If it is a leap year, state *two codes in their proper order*. Answers follow. No peeking. Use separate paper.

     1974   1996   1845   2019   2014   2016

Answers to *year code* problems:

1974

* 1) $74 \div 4 = 18$ which you reduce: $18 \equiv 4$ *(mod 7)*
* 2) $74 \equiv 4$ (mod 7)
* 3) The code for the 20th century is 5.
* 4) The sum of $4 + 4 + 5 = 13 \equiv 6$ (mod 7). The code for 1974 is 6.

1996 is a leap year

* 1) $96 \div 4 = 24 \equiv 3$ (mod 7)
* 2) $96 \equiv 5$ (mod 7)
* 3) The code for the 20th century is 5.
* 4) $3 + 5 + 5 = 13 \equiv 6$ (mod 7)

    6 is the *second code. The first code* is $6 - 1 \equiv 5$ (mod 7).

    The dual codes for 1996 *are 5 and 6.*

1845

* 1) $45 \div 4 = 11 \equiv 4$ (mod 7)
* 2) $45 \equiv 3$ (mod 7)
* 3) The code for the 19th century is 0.
* 4) $4 + 3 + 0 = 7 \equiv 0$ (mod 7)

    The code for 1845 is 0.

Observe that *the code for 1845 is 5 less than that for 1945.*

2019

* 1) $19 \div 4 = 4$
* 2) $19 \equiv 5$ (mod 7)
* 3) The code for the 21st century is 4.
* 4) The sum of $4 + 5 + 4 = 13 \equiv 6$ (mod 7). The code for 2019 is 6.

2014

* 1) $14 \div 4 = 3$
* 2) $14 \equiv 0$ (mod 7)
* 3) The code for the 21st century is 4.
* 4) The sum of $3 + 0 + 4 = 7 \equiv 0$ (mod 7). The code for 2014 is 0.

2016 is *a leap year*
- ❖ 1)　$16 \div 4 = 4$ (note: there is no remainder!)
- ❖ 2)　$16 \equiv 2 \pmod 7$
- ❖ 3)　The code for the 21st century is 4.
- ❖ 4)　The sum of $4 + 2 + 4 = 10 \equiv 3 \pmod 7$.

　　　　The *second code* for 2016 is 3. The *first* code is $3 - 1 = 2$.

　　　　The dual codes for 2016 are 2 and 3

Be *very* careful when finding year codes. B*eware of leap years. Always find two codes for leap years.* You are responsible to know which code to use depending on the month in play.

If a date is in January or February in a leap year, *use the first code.*

The Codes of The Months and Their Origins

We will take the same path that the person who created this machinery and we will see everything he (or she) saw in revealing the codes for the months. Forgive me, but for the sake of simplicity, the creator will be referred to as "he".

He set out to design a scheme that would tell the day of the week of any date.
Since everybody has a birthday and is curious about the day of the week on which they were born, this is called the Birthday Problem.

By default, the 19th century started with a year whose code was 0.
- January 1, 1800 falls on a Wednesday which is *has the code of 4.*
  I am guessing that he had a complete calendar for 1800.

His plan was to *add three codes; one for the **month**, one for the **date** and one for the **year**.* The sum of these three codes in modulo 7 would tell *the code of the day of the week* for that date, using the sequence from 0 to 6. For example, if the sum of the three numbers is 5, that would indicate Thursday because the code for Thursday is 5.

Here is the equation. We will refer to it as *the date algorithm* to distinguish it from *the year code algorithm*. Learn it. Solved problems will keep the order the same: The month codes first, then the date code, the year code and finally, the code of the day of the week.

> ➤ ***The Month code + the date + the year Code ≡ The Day of the Week (mod 7)***

The date is Wednesday, January 1, 1800.
The month code is currently unknown, so it will be represented by x.
The code for the date would be *the date itself (reduced, if necessary in modulo 7).*

In the case of *January 1, the date is 1*. Reducing is unnecessary. He posited that the year code for 1800 is 0.

He needed the code for January. *He used the bulleted equation above.* *x for the* **month code. The date** is 1. **The year code** is 0. **Wednesday** has code 4. He needed to solve $x + 1 + 0 \equiv 4 \pmod 7 \rightarrow x + 1 \equiv 4 \pmod 7 \rightarrow x \equiv 3 \pmod 7$ ***Thus, the code for January is 3***.

- On to February. He knew that *February 1, 1800 was a Saturday,* whose code is 0.

  $x + 1 + 0 \equiv 0 \pmod 7 \rightarrow x + 1 \equiv 0 \pmod 7 \rightarrow x \equiv \text{-}1 \equiv 6 \pmod 7$

Thus, the code for February is 6.

- March 1, 1800 *falls on a Saturday, too.*

Thus, the code for March is also 6.

- April 1, 1800 falls on a Tuesday, whose code is 3.

  $x + 1 + 0 \equiv 3 \pmod 7 \rightarrow x + 1 \equiv 3 \pmod 7 \rightarrow x \equiv 2 \pmod 7$

Thus, the code for April is 2.

- May 1, 1800 falls on a Thursday, whose code is 5.

  $x + 1 + 0 \equiv 5 \pmod 7 \rightarrow x + 1 \equiv 5 \pmod 7 \rightarrow x \equiv 4 \pmod 7$

Thus, the code for May is 4.

Note: No other month shares this code.

- June 1, 1800 falls on a Sunday, whose code is 1.

  $x + 1 + 0 \equiv 1 \pmod 7 \rightarrow x + 1 \equiv 1 \pmod 7$

Thus, the code for June is 0.

Note: No other month shares this code.

- July 1, 1800 falls on a Tuesday, whose code is 3.

  $x + 1 + 0 \equiv 3 \pmod 7 \rightarrow x + 1 \equiv 3 \rightarrow x \equiv 2 \pmod 7$

Thus, the code for July is 2.

Note: *The codes for April and July are <u>always</u> the same.*

- August 1, 1800 falls on a Friday, whose code is 6.

  $x + 1 + 0 \equiv 6 \pmod 7 \rightarrow x + 1 \equiv 6 \pmod 7 \rightarrow x \equiv 5 \pmod 7$

Thus, the code for August is 5.

Note: No other month shares this code.

- September 1, 1800 falls on a Monday, whose code is 2.

  $x + 1 + 0 \equiv 2 \pmod 7 \rightarrow x + 1 \equiv 2 \pmod 7 \rightarrow x \equiv 1 \pmod 7$

Thus, the code for September is 1.

- October 1, 1800 falls on a Wednesday, whose code is 4.

  $x + 1 + 0 \equiv 4 \pmod 7 \rightarrow x + 1 \equiv 4 \pmod 7 \rightarrow x \equiv 3 \pmod 7$

Thus, the code for October is 3.

- November 1, 1800 falls on a Saturday, whose code is 0.

  $x + 1 + 0 \equiv 0 \pmod 7 \rightarrow x + 1 \equiv 0 \pmod 7 \rightarrow x = -1 \equiv 6 \pmod 7$

Thus, the code for November is 6.

- December 1, 1800 falls on a Monday, whose code is 2.

  $x + 1 + 0 \equiv 2 \pmod 7 \rightarrow x + 1 \equiv 2 \rightarrow x \equiv 1 \pmod 7$

Thus, the code for December is 1.

Note: The codes for September and December are <u>always</u> the same.

Now we can show the codes in sequence. If you use this scheme in public, it is best if you memorize these codes for *effect*. BUT if you just look at 12 obscure numbers on a small piece of paper, people would still be impressed.

The month codes are:

Jan → (3)	Feb → (6)	Mar → (6)	Apr → (2)
May → (4)	June→ (0)	July → (2)	Aug → (5)
Sep → (1)	Oct → (3)	Nov → (6)	Dec → (1)

If you memorize them, group them in threes. 366 240 251 361

Friday the 13th

Before we proceed, there is a special talent that can be acquired with the knowledge of year codes. That talent is the ability to *find the month or months* in which the 13th will fall on a Friday, in other words, to find when Friday the 13th occurs in a specific year.

The only thing you must know is that every number in modulo 7 has a *complement*.
The *sum of a number and its complement is 0 in modulo 7*. We essentially covered complements with negative equivalents. *That means the two numbers must have a sum of 7.*

In the year in which this book was written, 2009, the year code is 1.
The complement of 1 in modulo 7 is 6 because 6 + 1 = 7.

The months that have *the code of 6* are February, March and November. Those are the months in 2009 in which Friday the 13th occurs in 2009. Warning: If this were a *leap year*, February would be segregated from March and November. If a leap year had the codes 5 and 6, for example, 1996, you would find the complement of the first code, 5.
The complement of 5 is 2 and neither January nor February have the code of 2. Thus, in January and February 1996, there is no Friday the 13th. However, in the rest of 1996, the code is 6 and its complement is 1. September and December have the code 1 and thus a Friday the 13th.

Find Friday the 13th in: 2013   2012   2011   2010
Try not to look at the answers.

The code for 2013 is 6. The complement of 6 is 1.

September and December both have the code 1.

Thus, September 13, 2013 and December 13, 2013 will both be Friday the 13th.

The dual codes for 2012 are (4, 5).

Using the *first code*, the complement of 4 is 3. January *has the code of 3*.

January 13, 2012 falls on Friday.

Using the second code, the complement of 5 is 2.

April and July share this code.

April 13, 2012 and July 13, 2012 fall on Friday.

Friday the 13th occurs three times in 2012.

The code for 2011 is 3. The complement of 3 is 4. May has the code of 4.

May 13, 2011 falls on Friday.

The dual code for 2016 is (2, 3). The complement of the first code, 2 is 5. Neither January nor February has that code. Moving on to the second code, the complement of 3 is 4.

The code of May is 4, thus May 13, 2016 falls on Friday.

TASK A: Solving The Birthday Problem

Here's a date: August 10, 1974. We will calculate the day of the week on which it falls. It is usually someone's birthday but this works for all dates in the Gregorian calendar.

We will use *the date algorithm*, introduced previously. Here are the ingredients.

1) You need the *month code*. Memorize the codes for effect.
2) You need *the date*.
3) You need *the code of the year*. You *may* have to calculate it.
   Do it on paper to be careful.
4) With practice, it can be done mentally. Find the year code *first*.
   Finally, add the month code, the date and the year code in modulo 7.

The *reduced* result will be *the code of the day of the week* on which the date falls. For August 10, 1974, we first **find the year code** for 1974 using the *year code algorithm*.

❖ a. $74 \div 4 = 18$ (drop the remainder!) Reduce $18 \equiv 4 \pmod 7$.
❖ b. $74 \equiv 4 \pmod 7$.
❖ c.   The code for the 20th century is 5.
❖ d.   Adding the three results $4 + 4 + 5 = 13 \equiv 6 \pmod 7$.
         Thus, the code for 1974 is 6.

Now that the year code is out of the way, we move on to the *date algorithm*.

▪ 1) The code for August is 5.
▪ 2) The date is 10 which you *can* reduce: $10 \equiv 3 \pmod 7$.
▪ 3) The code for 1974 is 6 (as we just calculated using the year code algorithm)
▪ 4. Add the three results: $5 + 3 + 6 = 14 \equiv 0 \pmod 7$.

The final result is 0 and that tells us that August 10, 1974 falls on a Saturday.

The month code + the date + the year code ≡ the day of the week (mod 7)
Whenever we use this equation, <u>one</u> of these four quantities will be unknown and represented by x.
In later chapters, the month, the date and the year will take turns being x in the equation.

After a while, you will remember the codes of *some* years and that will enable you to jump to *other* years. Speaking for myself, the codes of 1969, 1975, 1980, 1984, 1990, 1996, 1997, 1998, 1999, and 2003 are favorites. Their codes are all 0. I am particularly fond of "0 years" which I call years with *a code of 0*, as they give me a place to *start counting up*. It's also convenient to know that years whose last two digits are *divisible by 28* are leap years with codes 4, 5 in the 20th century and are leap years with the codes 3, 4 in the 21st century. This is because *28 is both divisible by 4 and 7*. Note that their second codes are the codes for their respective centuries.

It is *highly recommended* that you find the year code *first* to get it out of the way. Find it either by counting or by using the year code algorithm. The month codes are easy, especially if you memorized them. The date *may* need reducing. *Therefore, most of the messy work is in finding the year code.*

The current year is 2009 and the code is 1.
We will verify that using the *year code algorithm*.

- 09 ÷ 4 = 2 (and 2 does not require reducing because it is on the modulo 7 clock)
- 09 ≡ 2 (mod 7)
- The code for the 21st century is 4.
- Adding the 3 results: 2 + 2 + 4 ≡ 8 ≡ 1 (mod 7).

Thus, *the code of 1 is verified for 2009.*

What is the code for 2010? 2011? Do you need the algorithm?
What are the two codes for 2012?
What are the codes for 2006, 2007 and 2008?

If you move around the years with agility, you can avoid the year code algorithm. From a year *whose code you know*, all you have to do is add 1 to get the code of the next *year* (or the first code of a leap year). Just remember: leap years are going to come up. Beware of leap years.

If adding 1 to a year code gives you the code of a leap year, remember that is *only the first code*. The first code + 1 = the second code.
Distinguish between counting to get a year code for a leap year from using the year algorithm. If you count, you find the first code then add 1 to find the second code. If you use the algorithm, you find the second code THEN you SUBTRACT 1 to find the *first* code.

Did you find these codes? 2010(2) 2011(3) 2008(6, 0) 2007(5) 2006(4)
The following will show how you count from a year *you know*. We will count up. Counting backwards is a skill that is acquired and that is NOT necessary (or recommended).

The code for 2005 is 3.
Then add 1 to 3 and you have the code for 2006 which is 4.
Add 1 to 4 and you have the code for 2007 which is 5.
Add 1 to 5 and you have *the first code of* 2008 which is 6. Pause.
2008 is a leap year.
Add 1 and you have the *second* code for 2008 which is 0.

So 2008 *has two codes and we found them. The codes of 2008 are 6 and 0.*
If you continue adding 1, you *must resume from the second code of 2008 which is 0.*
Add 1 to 0 and you have the code for 2009, which is 1.
You can do this *indefinitely*.

Two solved problems will help you understand the date algorithm.

Example 1. On what day of the week does November 6, 1936 fall?

First, we will calculate the *dual codes* for 1936 because It is a leap year.
* 36 ÷ 4 = 9 There is no remainder because this is a leap year.
* $36 \equiv 1 \pmod 7$
* The century code is 5, for the 20th century
* The sum of the three numbers is $9 + 1 + 5 \equiv 15 \equiv 1 \pmod 7$.

Thus, the second code for 1936 is 1. We do not need the first code which is 0 because we are not dealing with January or February.

Now we use the *date algorithm*.
* The month code is 6.
* The date is 6, which does not require reducing.
* The year code is 1, as calculated using the year code algorithm.
* The sum of the three numbers is $6 + 6 + 1 \equiv 13 \equiv 6 \pmod 7$.

6 is the code for Friday.

That is the day of the week on which November 6, 1936 falls.

Example 2. On what day of the week does August 10, 1965 fall?

First, the year code algorithm will be used to calculate the code for 1965.
* 65 ÷ 4 = 16 Reduce $16 \equiv 2 \pmod 7$
* $65 \equiv 2 \pmod 7$
* The code for the 20th century is 5.
* The sum of the three numbers is $2 + 2 + 5 \equiv 9 \equiv 2 \pmod 7$.

Thus, the code for 1965 is 2.

Now we use the *date algorithm*.
* The month code is 5.
* The date is $10 \equiv 3 \pmod 7$.
* The year code is 2, as calculated using the year code algorithm.
* The sum of the three numbers is $5 + 3 + 2 \equiv 10 \equiv 3 \pmod 7$.

Thus the code for the day of the week is 3. August 10, 1965 falls on a Tuesday.

Task A:
Practice in Finding the Day of the Week of a DATE or the Birthday Problem

Find the day of the week for each date below. If the date in a *leap year*, *you* must determine *which code to use* based on the *month*.

1) March 17, 2006

2) February 29, 2008

3) September 13, 2010

4) December 13, 2010

5) May 31, 2005

6) July 14, 2009

7) October 27, 2011

8) April 12, 2007

9) March 1, 2008

10) December 21, 2012 (the day the *Mayan calendar predicts the world will end*)

Answers: Dates that are 7 or higher are *not reduced.*

1) March 17, 2006 falls on a Friday. $6 + 17 + 4 \equiv 27 \equiv 6 \pmod 7$

2) February 29, 2008 falls on a Friday. $6 + 29 + 6 \equiv 41 \equiv 6 \pmod 7$
   Did you use the right code for 2008?

3) September 13, 2010 falls on a Monday. $1 + 13 + 2 \equiv 16 \equiv 2 \pmod 7$

4) December 13, 2010 falls on a Monday. $1 + 13 + 2 \equiv 16 \equiv 2 \pmod 7$

5) May 31, 2005 falls on a Tuesday. $4 + 31 + 3 \equiv 38 \equiv 3 \pmod 7$

6) July 14, 2009 falls on a Tuesday. $2 + 14 + 1 \equiv 17 \equiv 3 \pmod 7$

7) October 27, 2011 falls on a Thursday. $3 + 27 + 3 \equiv 33 \equiv 5 \pmod 7$

8) April 12, 2007 falls on a Thursday. $2 + 12 + 5 \equiv 19 \equiv 5 \pmod 7$

9) March 1, 2008 falls on a Saturday. See problem (2) $6 + 1 + 0 \equiv 0 \pmod 7$

10) December 21, 2012 falls on a Friday. $1 + 21 + 5 \equiv 6 \pmod 7$

If you got 6 correct, rate yourself as passing in skill (D).
If you got 7 or 8 correct, rate yourself as satisfactory in skill. (C)
If you got 8 or 9 correct, rate yourself as superior in skill. (B)
If you got the all correct, rate yourself as an expert in Task A

Task A: Another test

Here are 20 more dates

1)   June 21, 2008

2)   November 30, 2010

3)   March 20, 2005

4)   December 31, 2009

5)   October 31, 2011

6)   May 28, 2006

7)   November 17, 2007

8)   July 29, 2007

9)   January 30, 2008

10)   March 21, 2008

11)   April 25, 2005

12)   August 5, 2009

13)   October 28, 2006

14)   February 28, 2007

15)   May 20, 2005

16)   July 12, 2008

17)   March 2, 2006

18)   June 17, 2005

19)   November 15, 2010

20)   20) April 15, 2007

Dates are reduced when possible.

1) June 21, 2008 falls on a Saturday. $0 + 0 + 0 \equiv 0$ (mod 7)

2) November 30, 2010 falls on a Tuesday. $6 + 2 + 2 \equiv 3$ (mod 7)

3) March 20, 2005 falls on a Sunday. $6 + 6 + 3 \equiv 1$ (mod 7)

4) December 31, 2009 falls on a Thursday. $1 + 3 + 1 \equiv 5$ (mod 7)

5) October 31, 2011 falls on a Monday. $3 + 3 + 3 \equiv 2$ (mod 7)

6) May 28, 2006 falls on a Sunday. $4 + 0 + 4 \equiv 1$ (mod 7)

7) November 17, 2007 falls on a Saturday. $6 + 3 + 5 \equiv 0$ (mod 7)

8) July 29, 2007 falls on a Sunday. $2 + 1 + 5 \equiv 1$ (mod 7)

9) January 30, 2008 falls on a Wednesday. $3 + 2 + 6 \equiv 4$ (mod 7) (*use the first code!*)

10) March 21, 2008 falls on a Friday. $6 + 0 + 0 \equiv 6$ (mod 7)

11) April 25, 2005 falls on a Monday. $2 + 4 + 3 \equiv 2$ (mod 7)

12) August 5, 2009 falls on a Wednesday. $5 + 5 + 1 \equiv 4$ (mod 7)

13) October 28, 2006 falls on a Saturday. $3 + 0 + 4 \equiv 0$ (mod 7)

14) February 28, 2007 falls on a Wednesday. $6 + 0 + 5 \equiv 4$ (mod 7)

15) May 20, 2005 falls on a Friday. $4 + 6 + 3 \equiv 6$ (mod 7)

16) July 12, 2008 falls on a Saturday. $2 + 5 + 0 \equiv 0$ (mod 7)

17) March 2, 2006 falls on a Thursday. $6 + 2 + 4 \equiv 5$ (mod 7)

18) June 17, 2005 falls on a Friday. $0 + 3 + 3 \equiv 6$ (mod 7)

19) November 15, 2010 falls on a Monday. $6 + 1 + 2 \equiv 2$ (mod 7)

20) April 15, 2007 falls on a Sunday. $2 + 1 + 5 \equiv 1$ (mod 7)

If you got 13-14 correct, rate yourself as *minimally competent. (C)*
If you got 15-16 correct, rate yourself as *satisfactory (B).*
If you got 17-18 correct, rate yourself as *superior (A).*
If you got 19-20 correct <u>you</u> are *excellent! (A+)*

Task B:
Find dates in a *given month* in a *given year* that fall on a *given day of the week.* or *The Appointment Calendar Problem*

The month code + the date (mod 7) + the year code $\equiv$ the day of the week

In this task, the <u>date</u> will be unknown. We will represent the date by x.

Suppose someone needs to meet with you in March 2010 and the meeting can *only* take place on a Thursday, whose code is 5.
We already know that the day of the week on which a date falls depends upon the month, the date and the code for the year. In this case, *the date is unknown.*

Let's build an equation. The *code for March is 6*. The date is x.
The *code for 2010 is 2.*
The code of the day of the week is *known* to be *5 for Thursday.*
Thus, the equation is $6 + x + 2 \equiv 5 \pmod 7$.

So we proceed: $6 + x + 2 \equiv 5 \pmod 7 \rightarrow x + 8 \equiv 5 \pmod 7 \rightarrow x = -3 \equiv 4 \pmod 7$.
The result is 4 and it tells us that *March **4**, 2010* falls on a Thursday.
So do March 11, 18 and 25. What if March 4, 2010 is a holiday and that person will be away? No problem. Make it March 11 or any other of the dates that fall on a Thursday. Try other solutions of the equation.

Feel free to refer to this example to solve the following problems. Some of these have traps. Don't fall into them. Be careful.

1. What dates in July, 2007 fall on a Saturday?
2. What dates in February 2008 fall on a Tuesday?
3. What dates in May 2009 fall on a Monday?
4. What dates in August 2005 fall on a Sunday?
5. What dates in November 2009 fall on a Sunday?

These questions are to test your earlier skills. The last two are hard.

6. On what day of the week did Christmas fall in 2007?
7. On what day of the week did the Fourth of July fall in 2005?
8. On what day of the week did New Year's Day fall in 2008? (compare to #6)
9. Labor Day falls on the *first Monday in September*. On what date does Labor Day fall in 2011? (Don't look at a calendar!)
10. Thanksgiving Day falls on the *last Thursday in November*. On what date does Thanksgiving Day fall in 2007?

Solved problems

1) What dates in July, 2007 fall on a Saturday?
   - The code for July is 2.
   - The date is x.
   - The code for 2007 is 5.
   - The code for Saturday is 0.

$2 + x + 5 \equiv 0 \pmod 7 \rightarrow x + 7 \equiv 0 \pmod 7 \rightarrow x \equiv 0 \pmod 7$

There is no such date as 0, so we *use* 7 instead.

Thus, July 7, 14, 21 and 28 fall on Saturday in 2007.

2) What dates in February 2008 fall on a Tuesday?
   - The code for February is 6.
   - The date is x.
   - The codes for 2008 are *6 and 0*.
     We must use the *first* code of 6 for February.
   - The code for Tuesday is 3.

$6 + x + 6 \equiv 3 \pmod 7 \rightarrow x + 12 \equiv 3 \pmod 7 \rightarrow x + 5 \equiv 3 \pmod 7 \rightarrow x = -2 \equiv 5 \pmod 7$

Thus, February 5, 12, 19 and 26 fall on Tuesday in 2008.

3) What dates in May 2009 fall on a Monday?
   - The code for May is 4.
   - The date is x.
   - The code for 2009 is 1.
   - The code for Monday is 2.

$4 + x + 1 \equiv 2 \pmod 7 \rightarrow x + 5 \equiv 2 \pmod 7 \rightarrow x = -3 \equiv 4 \pmod 7$

Thus, May 4, 11, 18 and 25 fall on Monday in 2009.

4) What dates in August 2005 fall on a Sunday?
   - The code for August is 5.
   - The date is x.
   - The code for 2005 is 3.
   - The code for Sunday is 1.

$5 + x + 3 \equiv 1 \pmod 7 \rightarrow x + 8 \equiv 1 \pmod 7 \rightarrow x + 1 \equiv 1 \pmod 7 \rightarrow x \equiv 0 \pmod 7$

The 0 is replaced with 7. Thus, August 7, 14, 21 and 28 fall on Sunday in 2005.

5) What dates in November 2009 fall on a Sunday?
- The code for November is 6.
- The date is x.
- The code for 2009 is 1.
- The code for Sunday is 1.

$6 + x + 1 \equiv 1 \pmod 7 \rightarrow x + 7 \equiv 1 \pmod 7 \rightarrow x + 0 \equiv 1 \pmod 7 \rightarrow x \equiv 1 \pmod 7$

Thus, November 1, 8, 15, 22 and 29 fall on a Sunday in 2009.

6) On what day of the week did Christmas fall in 2007?
(Christmas occurs on December 25th.)
- The code for December is 1.
- The date is 25 and $25 \equiv 4 \pmod 7$.
- The code for 2007 is 5.
- $1 + 4 + 5 = 10 \equiv 3 \pmod 7$ so Christmas day in 2007 falls on a Tuesday.

7) On what day of the week did the Fourth of July fall in 2005?
July 4, 2005 is the date in question.
- The code for July is 2.
- The date is 4.
- The code for 2005 is 3.
- $2 + 4 + 3 = 9 \equiv 2 \pmod 7$ so July 4, 2005 falls on a Monday.

8) On what day of the week did New Year's Day fall in 2008?
(compare answer to #6)
   New Year's Day is January 1.
- The code for January is 3.
- The date is 1.
- The dual codes for 2008 are 6 and 0. For January, we use the *first code*, 6.
- $3 + 1 + 6 = 10 \equiv 3 \pmod 7$ so January 1, 2008 falls on a Tuesday.

9) Labor Day falls on the *first Monday in September*.
   On what date does Labor Day fall in 2011?
- The code for September is 1.
- The date is x.
- The code for 2011 is 3.
- Monday has the code of 2.

$1 + x + 3 \equiv 2 \pmod 7 \rightarrow x + 4 \equiv 2 \pmod 7 \rightarrow x \equiv -2 \equiv 5 \pmod 7$

September 5, 2011 is the first Monday in September.

10) Thanksgiving Day falls on the *last Thursday in November*. On what date
does Thanksgiving Day fall in 2007?
- The code for November is 6.
- The date is x.
- The code for 2007 is 5.
- Thursday has the code of 5.

$6 + x + 5 \equiv 5 \pmod 7 \rightarrow x + 11 \equiv 5 \pmod 7 \rightarrow x + 4 \equiv 5 \pmod 7 \rightarrow x \equiv 1 \pmod 7$

Thus, November 1, 8, 15, 22 and 29 *all* fall on Thursday.

November 29, 2007 is Thanksgiving Day.

If you got 1 wrong, give yourself an A.

If you got 2 wrong, give yourself a B.

If you got 3 wrong, give yourself a C.

If you got 4 wrong, give yourself a D.

If you got all of them wrong, give yourself a flogging.

At this point, I expect that you are able to perform Task A and Task B with
the minimum of notes at hand. Beginners usually have to have the numbers
366240251361 handy (the codes of the months) or the codes of the *neighboring*
years, that is, 2 or 3 years ago or 2 or 3 years into the future. Experts will do year
codes quickly (and mentally) and therefore can ask for *any date in any century*.

If you are a very good at this, *all* calculations *can* be done mentally, even the
year codes. With experience, many year codes will stick in your memory.
Let's take a quick review of the system.

The 12 month codes are *memorizable*. It's almost like memorizing a phone
number.

The dates take care of themselves. Tip: If you are trying to find the day of
the week of a date, you should reduce the date *immediately* in mod 7
(if possible). Also, *as you work*, reduce *any* numbers.

Also, make substitutions. Substitute -1 for when adding 6, for example.

Year codes *can* be annoying, but if you practice, the *year code algorithm* will
be easy. After all, it only requires dividing a two-digit number by 4 and

reducing the same two-digit number in modulo 7. Knowing the multiples of 7 will give you an edge.

Also remembering some year codes will be useful.

Let's review the year algorithm and the date algorithm.

Example: April 26, 2042:

The code for 2042 is calculated first with the *year algorithm*:

- $42 \div 4 = 10 \equiv 3 \pmod 7$
- $42 \equiv 0 \pmod 7$
- The code for 21st century is 4
- $3 + 0 + 4 \equiv 7 \equiv 0 \pmod 7$

So *the code for 2042 is 0.*

Then on to the *date algorithm*:

- The code for April is 2
- The date is $26 \equiv 5 \pmod 7$
- The code for 2042 is 0.
- $2 + 5 + 0 \equiv 0 \pmod 7$. The result 0 tells us that *April 2, 2042 falls on a Saturday.*

All of these steps can be done mentally.

If you calculate the year code, keep the two algorithms separate. Focus. If you *know* the year code (or have it on a piece of paper) just perform the main (and easier) date algorithm. The mystery of how you do it will not be lost even if you take a few extra seconds calculating the year code.

Task C:
Finding YEARS with a Calendar Identical
to a given year or re-using calendars

With the Great Recession, all things that you would ordinarily throw out look more valuable.

Take your calendar. Do you want to throw it out on January 1? Years repeat their calendars in variable intervals and *leap years* repeat their calendars every 28 years.

There are three kinds of years between two consecutive leap years. Remember, a leap year is recognizable by the fact that the last two digits is a *multiple of a four*. The table in the book shows those multiples.

A *Leap Year +1* is the first type of year we will study and we will abbreviate it with *LY+1*.

An LY+1 year is a year that *follows* a leap year, that is, a leap year + 1.

LY+2 and LY+3 years are *2 and 3 years after a leap year, respectively*.

For example, 1985 and 1989 are LY+1 years. 1986 and 1990 are LY+2 years. 1987 and 1991 are examples of LY+3 years. It won't be critical to identify LY+2 and LY+3 years. But it *will be critical* to identify an LY+1 year and it *will be critical* to identify a *leap year*.

Rule 1. To find a year *identical* to an LY+1 year, *add 6 years to the year*.

Rule 2. To find a year *identical* to an LY+2 year *or* an LY+3 year, *add 11 years to the year*.

Suppose you want to reuse the calendar for 2005.
After you verify that the year is indeed an LY+1 year, *add 6* to 2005.
2005 + 6 = 2011. Thus 2011 is *identical* to 2005.
2005 and 2011 have *the same year code*. Verify that common year code.

Are you wondering if you can *repeat* that addition of 6 on 2011?
First, can you verify that 2011 is an LY+1? *It is not.* In fact, it's an LY+3 year.
You can only add 6 once to find one year that is identical to an LY+1 year.

What about repeating the addition of 11 on an LY+2 or LY+3 year?
1994 is an LY+2 year.
Adding 11 tells us that 2005 has the same calendar (and code) as 1994.
Can we add 11 again? 2005 is an LY+1 year.
We must add 6 to get the next identical year. 2005 + 6 = 2011 and thus
2011 is identical to 2005 and 1994. Find their common code.

1995 is an LY+3 year.
Adding 11 tells us that 2006 has the same calendar (and code) as 1995.
Can we add 11 again? 2006 is an LY+2 year, *so yes we can.* 2006 + 11 = 2017.
Thus 1995, 2006 and 2017 have identical calendars.
Find their common code.

Exercises:

1) Find the *first year* after 1945 with the same calendar.
2) Find the *first year* after 1959 with the same calendar.
3) Find the *first year* after 2018 with the same calendar.

Answers

1) 1945 is an LY+1. 1945 + 6 = 1951. 1945 and 1951 share the code of 5.
2) 1959 is an LY+3. 1959 + 11 = 1970. 1959 and 1970 share the code of 1.
3) 2018 is an LY+2. 2018 + 11 = 2029. 2018 and 2029 share the code of 5.

More on LY+2 and LY+3 years

o   Example 1. 1978 is an example of an LY+2. Add 11 years to reach a year with
the same calendar. 1978 + 11 = 1989. 1978 and 1989 have the same calendar.
Can you add 11 to 1989 again and find *the next year* with the same calendar?

1989 is an LY+1 year and you would *add 6* to get the next identical year!
1989 + 6 = 1995 which is *identical* to 1989.
Then 1978, 1989 and 1995 all have *identical calendars*. Find their common code.
So once again, the addition of 11 years can *not* always be repeated.

o   Example 2. 1987 is an LY+3 year.
1987 + 11 = 1998. The calendars for 1987 and 1998 are identical.
1998 is an LY+2 so we *can add 11* again! 1998 + 11 = 2009.
2009 also has the same calendar.
Find the common code for 1987, 1998 and 2009.

For the mathematically inclined, here is the reason for these phenomena.
Skip this if math gives you a headache, It is not essential to know it or even to
understand it. If it gives you a headache, skip it and move on to the exercises.
Leap years are in the form 4k, where k is any positive integer.

4k + 3 represents an LY+3 year. 4k + 3 + 11 represents the *first* addition of 11.
That expression simplifies to 4k + 14 = 4(k + 3) + 2 which is an LY+2 year.

4(k + 3) + 2 represents *an LY+2 year (because it shows addition of 2)*.
Then we can add 11 *again*.
We return to its original form, 4k + 14 and change to 4k + 12 + 2.

However, if you start out with an LY+2 year which *is represented by 4k + 2*,
and add 11, you will get 4k + 2 + 11 = 4k + 13 = 4k + 12 + 1 = 4(k + 3) + 1,
which represents an LY+1 year!
Then you would add 6.

When you start with an LY+3 year and add 11, you will get an LY+2 year. Then you can add 11 *again*.

When you start with an LY+1 year and add 6, you end up with an *LY+3 year*.
Under no circumstances can you perform two successive additions of 6.
In example 1, the years are 1978, 1989 and 1995. Their codes are all 4.
In example 2, the years are 1987, 1998 and 2009. Their codes are all 1.

Exercises: Find the next *two* years with the same calendar as each of the following.
1) 2002   2) 2015   3) 2009

Answers:

1)   2002 is an LY+2 year thus, 2002 + 11 = 2013, which is an LY+1 year. 2013 + 6 = 2019. 2002, 2013 and 2019 share the common code 6.

2)   2015 is an LY+3 year thus, 2015 + 11 = 2026, which is an LY+2 year. Thus, 2026 + 11 = 2037. 2015, 2026 and 2037 share the common code 1. (Note: In this example, we were able to add 11 twice in succession.)

3)   2009 is an LY+1 year thus 2009 + 6 = 2015, which is an LY+3 year. Thus, 2015 + 11 = 2026. 2009, 2015 and 2026 share the common code 1. Notice the overlapping of the years 2015 and 2026 in the answers.

First, it is important to add a word about the last decade of the 19th century. Since 1900 is not a LEAP YEAR, the years beginning with 1889 to 1900 do *not conform* to the rules described above. *From 1889 to 1900*, in order to find a year with the same calendar as any one of them (except leap years), 12 must be added to the year *regardless* of the nature of the year (LY+2 or LY+3). 1889 and 1893 are the most strange. They are LY+1 years. You can add 6 <u>or</u> 12 to them to find years with the same calendar!
1893, 1893 + 6 = 1899 and 1883 + 12 = 1905 *all have the code 4*. 1900, *as the last year of the decade*, also behaves strangely.*1900 + 6* is the *only* way to acquire a year with same calendar.
Starting with 1901, the *first year of the 20th century*, things settle down and you can relax until 2090 when the same circus visits town.
This is what happens in shifting from one *non-leap century* to the *next non-leap century*.
Thankfully, that won't happen again until the decade of 2090 to 2099.

Find the *next two years* (and codes) after the given year that has the *same* calendar.

1) 1994

2) 1935

3) 1954

4) 2001

5) 1922

6) 1943

7) 2003

8) 1999

9) 1914

10) 1827

A TIP: If you get a *LEAP YEAR* as an answer, STOP! That is *impossible!*

Answers:

1) 1994 is an LY+2 year thus, 1994 + 11 = 2005, which is an LY+1 year. 2005 + 6 = 2011. The years 1994, 2005 and 2011 are identical. Their common code is 3

2) 1935 is an LY+3 year thus, 1935 + 11 = 1946, which is an LY+2 year. 1946 + 11 = 1957. The years 1935, 1946 and 1957 are identical. Their common code is 6.
Note: We were able to perform the repeated addition of 11.

3) 1954 is an LY+2 year thus, 1954 + 11 = 1965, which is an LY+1 year. 1965 + 6 = 1971. The years 1954, 1965 and 1971 are identical. Their common code is 2.

4) 2001 is an LY+1 year thus, 2001 + 6 = 2007, which is an LY+3 year.
   2007 + 11 = 2018. The years 2001, 2007 and 2018 are identical.
   Their common code is 5.

5) 1922 is an LY+2 year thus, 1922 + 11 = 1933, which is an LY+1 year.
   1933 + 6 = 1939. The years 1922, 1933 and 1939 are identical.
   Their common code is 4.

6) 1943 is an LY+3 year thus, 1943 + 11 = 1954, which is an LY+2 year.
   1954 + 11 = 1965. The years 1943, 1954 and 1965 are identical.
   Their common code is 2.
   Note: We were able to perform the repeated addition of 11.

7) 2003 is an LY+3 year thus 2003 + 11 = 2014, which is an LY+2 year.
   2014 + 11 = 2025. The years 2003, 2014 and 2025 are identical.
   Their common code is 0.
   Note: We were able to perform the repeated addition of 11.

8) 1999 is an LY+3 year thus 1999 + 11 = 2010, which is an LY+2 year.
   2010 + 11 = 2021. The years 1999, 2010 and 2021 are identical.
   Their common code is 2.
   Note: We were able to perform the repeated addition of 11.

9) 1914 is an LY+2 year thus 1914 + 11 = 1925, which is an LY+1 year.
   1925 + 6 = 1931. The years 1914, 1925 and 1931 are identical.
   Their common code is 1.

10) 1827 is an LY+3 year thus 1827+ 11 = 1838, which is an LY+2 year.
    1838 + 11 = 1849. The years 1827, 1838 and 1849 are identical.
    Their common code is 5.
    Note: We were able to perform the repeated addition of 11.
    The century is irrelevant.

If you got them ALL RIGHT, give yourself an A+.
If you got 1 or 2 wrong, give yourself an A.
If you got 3 or 4 wrong, give yourself a B.
If you got 5 or 6 wrong, give yourself a C.
If you got 7 or 8 wrong, give yourself a D.
If you got 9 or 10 wrong, give yourself an F.

Task D: The Law & Order Problem

The month code + the date (mod 7) + the year code ≡ the day of the week

This is the most tedious of all the tasks. In this task, the *year* is unknown. Any fan of old Law & Order episodes is familiar with an occasional *deep musical chord* which signals the appearance of a *date* at the bottom of the screen. The date had all the ingredients of the *date algorithm except the year*. The date comes in this form: **Friday, September 29**. The task is to *find the year or years* in which the date *could* have occurred.

Let's find out in what (recent) years this Law & Order episode could have been made. This task is the most tedious.
We will examine *Friday, September 29*.

➢ September has the code 1
➢ The date is $29 \equiv 1 \pmod 7$
➢ The code of the year is unknown, so we will call it x.
➢ The code for the day of the week is 6, for Friday.

The date algorithm develops as $1 + 1 + x \equiv 6 \pmod 7 \rightarrow x + 2 \equiv 6 \pmod 7$ $x \equiv 4 \pmod 7$. Thus, the *code for the year was* 4. What *recent years* had the code of 4?

We will only use years in the 21ˢᵗ century and the late 20ᵗʰ century.
The game of *concentration* comes to mind. The same year codes appear over and over because they are on a (circular) modulo 7 clock. Remember the spinner. The last non-leap year that had a code of 4 in the 20ᵗʰ century was *1995*. This is a solution.

2000 has the *dual codes 3 and 4*. Here we also have another solution. September is in *that part of 2000 in which the code is 4.*

If you count up (with your fingers) from 2000, you can find another solution. Add 4 + 1 = 5 and we get the code for 2001. Add 5 + 1 = 6 and we get the code for 2002. Add 6 + 1 = 0 and we get the code for 2003.

No luck yet. Adding 1 to the code for 2003, we get 0 + 1 = 1, the first code of 2004, a leap year. The dual codes for 2004 are 1 and 2. We resume counting from the *second code*, 2.
Add 2 + 1 = 3, the code for 2005. Add 3 + 1 = 4 and bingo.
We found another solution.
4 is the code for 2006 and that tells us that 2006 is a second solution in the 21st century.
Friday, September 29 could have occurred in 2000 or 2006. Their year codes agree.
With practice, this moves more quickly.

Persistence is important.

Find years in which an episode of Law & Order *could* have been made. Restrict answers to the 21st century. Pay attention to the month. If it is January or February, it may be in a leap year!

    1) Tuesday, February 10   2) Wednesday, March 17   3) Saturday, July 15

    4) Monday, January 14   5) Thursday, October 28

Answers:

1) Tuesday, February 10
- February has the code 6.
- The date is $10 \equiv 3 \pmod 7$.
- The year is x.
- Tuesday has the code of 3.

$6 + 3 + x \equiv 3 \pmod 7 \rightarrow x + 9 \equiv 3 \pmod 7 \rightarrow x + 2 \equiv 3 \pmod 7 \rightarrow x \equiv 1 \pmod 7$

So the code of a year in which this episode *could have been made is* 1.

We are looking for *non-leap* years whose code is 1 *or leap years* whose <u>first code</u> is 1 because *the date precedes March.*

2004 has the dual code (1, 2) and *is the only possible leap year because the first code is 1.*

2009 is another solution. So 2004 and 2009 are the solutions.

2) Wednesday, March 17
- March has the code 6.
- The date is $17 \equiv 3 \pmod 7$.
- The year is x.
- The code for Wednesday is 4.

$6 + 3 + x \equiv 4 \pmod 7 \rightarrow x + 9 \equiv 4 \pmod 7 \rightarrow x + 2 \equiv 4 \pmod 7 \rightarrow x \equiv 2 \pmod 7$

The code of any possible year is 2.

If you remembered that 2003 has a code of 0, it's a quick hop to 2004.

2004 has codes of 1 and 2. For *March, the second code of 2 applies.*

So 2004 is a solution.

Are there any other recent (ordinary) years with code 2?

You have to go back to 1999.

The only recent solution is 2004.

3) Saturday, July 15
   - July has the code 2.
   - The date is $15 \equiv 1$ (mod 7).
   - The year is x.
   - The code for Saturday is 0.

   $2 + 1 + x \equiv 0$ (mod 7) $\rightarrow x + 3 \equiv 0$ (mod 7) $\rightarrow x \equiv -3 \equiv 4$ (mod (7)

   Thus, the code of any possible year is 4.

   Again, 2000 has the codes 3 and 4. The second code 4 works with July in 2000.

   Search for other code 4 years: Start with 2001 that has the code of 5.

   2001(5) 2002(6) 2003(0) 2004(1, 2) 2005(3) 2004(4) Stop.

   There are two solutions: 2000 and 2006.

4) Monday, January 14
   - January has the code 3.
   - The date is $14 \equiv 0$ (mod 7).
   - The year is x.
   - The code for Monday is 2.

   $3 + 0 + x \equiv 2$ (mod 7) $\rightarrow x + 3 \equiv 2$ (mod 7) $\rightarrow x \equiv -1 \equiv 6$ (mod 7)

   The code of any possible year is 6.

   We are looking for *non-leap years* with a code of 6 and *leap years* with a first code of 6.

   Look back at the previous solutions. 2002 qualifies.

   2008 has two codes 6 and 0.

   The *first code applies to January* and the date is *in January!*

   2002 and 2008 are the solutions.

5) Thursday, October 28
   - October has the code 3.
   - The date is $28 \equiv 0$ (mod 7).
   - The year is x.
   - The code for Thursday is 5.

   $3 + 0 + x \equiv 5$ (mod 7) $\rightarrow x + 3 \equiv 5$ (mod 7) $\rightarrow x \equiv 2$ (mod (7)

   The code of any possible year is 2.

   2004 has the codes 1 and 2. The second code fits October since October is after February.

   2010 has the code of 2 but it is doubtful if anyone would care about 2010 yet.

   It all depends on when this book is being read. Tempus fugit.

   So the only current solution is 2004.

Task E:
Find the month(s) in which a given date in a given year can fall on a given day of the week

In Task A, the day of the week of a date was unknown.
In Task B, the numerical date within a month was unknown.
In Task C, we found years in which a calendar can be re-used.
In Task D, the year of a date was unknown.
In Task E, the month (or <u>months</u> in some cases) is (are) unknown.

A possible scenario: you are a spy. Another agent tells you to meet him on Tuesday, the 10th of the month in 2013, but he does not tell *which* month. Warning: There may be *no solution* to some of these problems.

Working with the *date algorithm*:
- ✓ The code for the month is unknown, so we will call it x.
- ✓ The date is $10 \equiv 3 \pmod 7$.
- ✓ The code for 2013 is 6.
- ✓ *The day of the week is Tuesday, whose code is 3.*
    $x + 3 + 6 \equiv 3 \pmod 7 \rightarrow x + 9 = 3 \rightarrow x = -6 \equiv 1 \pmod 7$
The result is 1 which is the code for September and December.

Beware of pitfalls.
For example, what months will have the date Monday the 30th in 2009?
- ✓ The month is unknown, so we will call it x.
- ✓ The date is $30 \equiv 2 \pmod 7$.
- ✓ The year has the code 1.
- ✓ The code for Monday is 2.
Thus, $x + 2 + 1 \equiv 2 \pmod 7 \rightarrow x + 3 \equiv 2 \pmod 7 \rightarrow x \equiv -1 \equiv 6 \pmod 7$
February, March and November have the code of 6 in 2009.
However, February only has 28 days in 2009.
So the solutions are March and November.

Another example: The 31st falls on a Wednesday in 2009.
What month(s) are possible?
✓   The code for the month is unknown, so we will call it x.
✓   The date is $31 \equiv 3 \pmod 7$.
✓   The code for 2009 is 1.
✓   The day of the week is known to be Wednesday, whose code is 4.
$x + 3 + 1 \equiv 4 \pmod 7 \rightarrow x + 4 = 4 \rightarrow x \equiv 0 \pmod 7$
June is the only month with the code of 0. *But June only has 30 days!*
There is NO SOLUTION. Beware of pitfalls.

In every year, there is *at least* one date that is non-existent. In a *leap year*,
there are *two impossible dates*. More on that will follow.
In 2009, that non-existent date is Wednesday, the 31st. June is responsible.
Solve $x + 3 + 1 \equiv 4 \pmod 7$ and you will see why. Remember that x
represents the code of a month! When you match a month code to your
solution, you will see what I mean.

Find *a day of the week* that can NOT be on the 31st in 2008, 2007, 2006,
2005, 2004, 2003, 2002, 2001 and 2000.
There is a way to find those impossible dates quickly using the year code.
Because $31 \equiv 3 \pmod 7$, all you have to do is *add 3 to the year code*. This is because
June is responsible for those impossible dates and the code for June is 0.
Using 2011 as an example, its code is 3. The date algorithm has us adding:
the month code + the date + the year code $\equiv$ the code for the day of the week
(mod 7)

$$\underline{0 \ + \ 31} \qquad + \qquad 3 \qquad \equiv \qquad x$$

In this case, it becomes $0 + 31 + 3 \equiv$ the code for the day of the week (mod 7).
Reducing 31 to 3, this becomes $0 + 3 + 3 \equiv 6 \pmod 7$ which is the code for
Friday.

So in 2011, the 31st *cannot fall on a Friday*. This is all the fault of *June.*

June has only 30 days. If June had 31 days, then June 31, 2011 *would* fall on a Friday. To summarize, because the sum of the month code and the date code is 3, we *add 3* to the year code to find the code for the day of the week. For example, 2002 has the code of 6. Adding 6 + 3 gives us 9 ≡ 2 (mod 7). 2 is the code for Monday. Then the 31st cannot fall on a Monday in 2002.

For the mathematically curious, here is a breakdown of why this works. The month code 0 is unique. Only June has it and June only has 30 days. The month code 1 is *not* unique. December has that code and it has 31 days. The month code 2 is *not* unique. July has it and that code and has 31 days. The month code 3 is *not* unique.
January and October have that code and *both* have 31 days.
The month code 4 (for May) is unique, *but May has 31 days*.
The month code for August is 5 is unique but August has *31 days*.

February has the code of 6. February can also be used to find impossible dates. But you can do this only in *leap years*.
You know that leap years have dual codes.
To find dates that are impossible caused by February's deficit in days, just *add 2 to the first year code*. That will find a day of the week in that year that cannot fall on the 31st. This works because 6 + 3 = 9 which is equivalent to 2 in mod 7. If you *add 1 to the first year code of a leap year*, you will find a day of the week that cannot fall on the 30th. The first code for 2008 is 6.
Adding 6 + 2 = 8 ≡ 1 (mod 7), which means that the 31st cannot fall on a Sunday in 2008. If you add 6 + 1 = 7 ≡ 0 (mod 7), the 30th cannot fall on a Saturday in 2008. If you move to the second code, 0, you will add 3 as we did before, because now June is the guilty month. Adding 0 + 3 ≡ 3 (mod 7), then the 31st cannot fall on Tuesday in 2008.
So in 2008, Sunday the 31st, Saturday the 30th and Tuesday the 31st are impossible dates.

Let's explore 1984 whose codes are (4, 5).
Adding 4 + 2 ≡ 6 (mod 7), so Friday the 31st is impossible.
Adding 5 + 3 ≡ 8 ≡ 1 (mod 7), so Sunday the 31st is impossible.

To summarize, *in a leap year*, there are *at least two* dates that will be impossible.

To find the day of the week on which the 31st cannot fall, *add 2* to the *first code*. The result is caused by February. Then add 3 to the second code to find the other day of the week. This one is caused by the deficit of days in June.

In a *non-leap year*, there will be *only one* impossible day of the week that cannot fall on the 31st. That will be caused by June.

Add 3 to the year code to find it.

Problems: Find the month(s) in which the following dates occur. Solutions appear below.

1) The 8th falls on a Tuesday in 2001

2) The 23rd falls on a Thursday in 1997

3) The 21st falls on a Friday in 2000

4) The 18th falls on a Saturday in 2007

5) The 24th falls on a Thursday in 1998

6) The 31st falls on a Sunday in 2005

7) The 16th falls on a Wednesday in 1999

8) The 22nd falls on a Saturday in 2002

9) The 16th falls on a Friday in 1976

10) The 8th falls on a Sunday in 2006

Solutions:

1) The 8th falls on a Tuesday in 2001
   - ❖   Let x = the code of the month.
   - ❖   The date is $8 \equiv 1 \pmod 7$.
   - ❖   The code for 2001 is 5.
   - ❖   The code for Tuesday is 3.

   $x + 1 + 5 \equiv 3 \pmod 7 \rightarrow x + 6 \equiv 3 \pmod 7 \rightarrow x = -3 \equiv 4 \pmod 7$

   Then the code of the month is 4 which makes *May* the only solution.

2) The 23rd falls on a Thursday in 1997
   - ❖   Let x = the code of the month.
   - ❖   The date is $23 \equiv 2 \pmod 7$.
   - ❖   The code for 1997 is 0.
   - ❖   The code for Thursday is 5.

   $x + 2 + 0 \equiv 5 \pmod 7 \rightarrow x + 2 \equiv 5 \pmod 7 \rightarrow x \equiv 3 \pmod 7$

   Then the code of the month is 3 which makes *January and October* solutions.

3) The 21st falls on a Friday in 2000
   - ❖   Let x = the code of the month.
   - ❖   The date is $21 \equiv 0 \pmod 7$.
   - ❖   The dual codes for 2000 are *3 and 4*.
   - ❖   The code for Friday is 6.

   Try the *first code* of 3: $x + 0 + 3 \equiv 6 \pmod 7 \rightarrow x + 3 \equiv 6 \pmod 7 \rightarrow x \equiv 3 \pmod 7$
   Since *January is valid for the first code,* and its code is 3, we have a solution.
   Try the second code: $x + 0 + 4 \equiv 6 \pmod 7 \rightarrow x + 4 \equiv 6 \pmod 7 \rightarrow x \equiv 2 \pmod 7$
   The months that have a code of 2 *after February 29* are April and July.
   So the solutions are *January, April and July.*

4) The 18th falls on a Saturday in 2007
   - ❖   Let x = the code of the month.
   - ❖   The date is $18 \equiv 4 \pmod 7$.
   - ❖   The code for 2007 is 5.
   - ❖   The code for Saturday is 0.

   $x + 4 + 5 \equiv 0 \pmod 7 \rightarrow x + 9 \equiv 0 \pmod 7 \rightarrow x + 2 \equiv 0 \pmod 7 \rightarrow x \equiv -2 \pmod 7$
   $x \equiv 5 \pmod 7$, thus the code of the month is 5 which makes *August* the *only* solution.

5) The 24th falls on a Thursday in 1998
   ❖   Let x = the code of the month.
   ❖   The date is $24 \equiv 3$ (mod 7).
   ❖   The code for 1998 is 1.
   ❖   The code for Thursday is 5.

   $x + 3 + 1 \equiv 5$ (mod 7) $\rightarrow x + 4 \equiv 5$ (mod 7) $\rightarrow x \equiv 1$ (mod 7)

   Then the code of the month is 1 which makes *September and December* solutions.

6) The 31st falls on a Sunday in 2005
   ❖   Let x = the code of the month.
   ❖   The date is $31 \equiv 3$ (mod 7).
   ❖   The code for 2005 is 3.
   ❖   The code for Sunday is 1.

   $x + 3 + 3 \equiv 1$ (mod 7) $\rightarrow x + 6 \equiv 1$ (mod 7) $\rightarrow x = -5 \equiv 2$ (mod 7).

   Then the code of the month is 2 which makes *July* the only solution.
   April *also* has the code of 2, but it only has *30* days.

7) The 16th falls on a Wednesday in 1999
   ❖   Let x = the code of the month.
   ❖   The date is $16 \equiv 2$ (mod 7).
   ❖   The code for 1999 is 2.
   ❖   The code for Wednesday is 4.

   $x + 2 + 2 \equiv 4$ (mod 7) $\rightarrow x + 4 \equiv 4$ (mod 7) $\rightarrow x \equiv 0$ (mod 7)

   Then the code of the month is 0 which makes *June* the only solution.

8) The 22nd falls on a Saturday in 2002
   ❖   Let x = the code of the month.
   ❖   The date is $22 \equiv 1$ (mod 7).
   ❖   The code for 2002 is 6.
   ❖   The code for Saturday is 0.

   $x + 1 + 6 \equiv 0$ (mod 7) $\rightarrow x + 7 \equiv 0$ (mod 7) $\rightarrow x + 0 \equiv 0$ (mod 7) $\rightarrow x \equiv 0$ (mod 7)

   Then the code of the month is 0 which makes *June* the only solution.

9) The 16th falls on a Friday in 1976
- ❖   Let x = the code of the month.
- ❖   The date is $16 \equiv 2$ (mod 7).
- ❖   The dual codes for 1976 are 1 and 2.
- ❖   The code for Friday is 6.

Try the first code of 1:

$x + 2 + 1 \equiv 6$ (mod 7) $\rightarrow x + 3 \equiv 6$ (mod 7) $\rightarrow x \equiv 3$ (mod 7)

3 is the code for *January*. So we have one solution, January.

Try the second code of 2:

$x + 2 + 2 \equiv 6$ (mod 7) $\rightarrow x + 4 \equiv 6$ (mod 7) $\rightarrow x \equiv 2$ (mod 7).

The code of 2 applies to *April and July*.

Then there are *three* months in which the 16th falls on a Friday in 1976. *January, April and July.*

10) The 8th falls on a Sunday in 2006
- ❖   Let x = the code of the month.
- ❖   The date is $8 \equiv 1$ (mod 7).
- ❖   The code for 2006 is 4.
- ❖   The code for Sunday is 1.

$x + 1 + 4 \equiv 1$ (mod 7) $\rightarrow x + 5 \equiv 1$ (mod 7) $\rightarrow x = -4 \equiv 3$ (mod 7)

Then the code of the month is 3 which make *January and October* solutions.

These are the answers to the exercises suggested to find a day of the week for the years from 2000 to 2008 on which *the 31st can not* fall:

2000: Thursday and Saturday
2001: Sunday
2002: Monday
2003: Tuesday
2004: Tuesday and Thursday
2005: Thursday
2006: Saturday
2007: Sunday
2008: Sunday and Tuesday
The culprit is June, in all non-leap years.
In leap years, June and February share the blame.

Task F:
Reducing LARGE NUMBERS (3 and 4 digits) in Modulo 7

Finally, I'd like to share another skill.

A very simple task is to find the day of the week that is a certain number of days *after a given day of the week*. For example, what day of the week is 15 days after Tuesday?

We will use modulo 7 arithmetic. You *could* use the clock but many of these numbers are too large for that. You will get dizzy.

Starting at 3 which is the code for Tuesday, add $3 + 15 = 18 \equiv 4 \pmod 7$.

The result tells us that a 15 day advance from will end on *Wednesday, whose code is 4.*

Since $15 \equiv 1 \pmod 7$ means that a 15 day advance is equivalent to a 1 day advance! A 1 day advance from Tuesday ends on Wednesday. The answer is the same. You can do problems like these by reducing the number of days you want to advance.

Suppose today is Tuesday. In 489 days what will the day of the week be?
This is the algorithm for a 3 digit number.

1) Separate the hundred's place, which is 4, from 89.
   Double the hundred's place, and you get 8.
2) *Add* 8 to 89 and you get 97.
3) Now *reduce 97 in modulo 7*. $97 - 91 = 6 \equiv 6 \equiv 6 \pmod 7$.
   6 is the number of days of the week to ADVANCE from Tuesday.
   The code for Tuesday is 3 and $3 + 6 = 9 \equiv 2 \pmod 7$.
   So the result of 2 tells that the solution is *Monday because its code is 2.*

If you started by adding the code of Tuesday and 489 *without* reducing, this is what happens:

489 + 3 = 492. Separate 4 from 92, double 4 and add 8; 8 + 92 = 100 ≡ 2 (mod 7). The result still tells us that you will arrive on a Monday, whose code is 2.

There is a trick that will make these easier. It was mentioned earlier.
From left to right, you can reduce digits in modulo 7 that are more than 6.
These digits are 7, 8 and 9.
The underlined portion of the number is going to be changed using modulo 7.
4*89 becomes 4*12 in modulo 7. *One digit at a time can be reduced.*
This is because 8 ≡ 1 and 9 ≡ 2 (mod 7).
Then following the 3 steps above:

1) *Separate 4 from 12*
2) *Double 4. Add 8 to 12 and you get 20*
3) Reduce 20: 20≡ 6 *(mod 7).*

Or 4*89 becomes 6*9 ≡ 6 (mod 7) because 48 ≡ 6 (mod 7). Then 69 ≡ 6 (mod 7).
Or 4*89 becomes 4*05 (mod 7) because 89 ≡ 405 (mod 7).
(It is essential to keep the number of digits the same when you are not working from the left-most digit.)
Separate 4 from 05. Double 4. Add 8 + 05. The result is 13 ≡ 6 (mod 7).

No matter what you do, as long as it's a *correct action*, you get the same result.
One more example: What day of the week is *525* days after a Wednesday?
Wednesday is the 4th day of the week.

- Separate 5 from 25
- Double 5 and the result is 10.
- Add 10 to 25 and the result is 35
- *Reduce 35 in mod 7.*
- 35 ≡ 0 (mod 7) because 35 is a multiple of 7.

No days will advance because of the 0 result.

Explore this problem. Solve it in as many ways as possible.

To find the day of the week 3,184 days from today, there is another *algorithm for four-digit numbers*

- Separate *the thousand's place, 3 from 184*
- Subtract 3 from 184. You get 181. There is no doubling here.

Now apply the previous algorithm for a three digit number.

- Separate 1 from 81
- Double 1 and you get 2
- Add 2 to 81 and you get 83
- *Reduce 83 ≡ 6 (mod 7).*

Thus, we advance 6 days from Wednesday and arrive at *Tuesday.*

Note: 6 ≡ -1 (mod 7) so you could have moved *BACK 1 day* instead.

The main value of this is to exercise your knowledge of modulo 7.

It also can useful to find if a number is (or is not) a multiple of 7.

Here are some examples of shortcuts in reducing large numbers using modulo 7.

14,101 can be whittled down by deleting 14 from the left.

101 is the number now.

Now you will separate 1 from 01. Double 1 and add the result to 01.

2 + 1 = 3. Reducing is unnecessary.

Replacing digits *from the right*, or any digit *other than the first* digit, you have to be careful.

In reducing 927, the 7 can be replaced by 0 in modulo 7.

This changes 927 to 920.

92 can be reduced to 1. Then the number becomes 10 ≡ 3 (mod 7).

We are done. There is an advance of 3 days. It doesn't matter how you do it.

Or the 27 can be changed to 06 creating 906. And the 9 can be changed to 2.

206 can now be reduced easily in (mod 7).

Separate 2 from 06, double 2 and add 4 + 6 ≡ 10 ≡ 3 *(mod 7). It's mental acrobatics.*

Double, add and reduce. That's all you have to know to reduce a three-digit number.

Reduce the following numbers. Don't look at the answers below.

1) 639 ≡   (mod 7) 639 days after Wednesday is _____
2) 299 ≡   (mod 7) 299 days after Saturday is _____
3) 756 ≡   (mod 7) 756 days after Friday is _____
4) 3,059 ≡   (mod 7) 3,056 days after Saturday is _____
5) 8,480 ≡   (mod 7) 8,480 days after Thursday is _____
6) 9,253 ≡   (mod 7) 9,253 days after Tuesday is _____
7) 5,902 ≡   (mod 7) 5,902 days after Sunday is _____
8) 6,006 ≡   (mod 7) 6,006 days after Wednesday is _____
9) 5,005 ≡   (mod 7) 5,005 days after Tuesday is _____
10) 4,005 ≡   (mod 7) 4005 days after Thursday is _____

Answers: Steps in the algorithm(s) are compacted. Not all possible solutions are shown. No shortcuts were taken. That is left up to you. Experiment.

1) 639
    ❖ Doubling 6 yields 12. Add: 12 + 39 = 51
    ❖ 51 ≡ 2 (mod 7) thus 2 days after Wednesday is *Friday: 4 + 2 = 6*

2) 299
    ❖ Doubling 2 yields 4. Add: 4 + 99 = 103
    ❖ 103 ≡ 5 (mod 7) and 5 days after Saturday is *Thursday: 0 + 5 = 5*

3) 756
    ❖ Doubling 7 yields 14. Add 14 + 56 = 70
    ❖ 70 ≡ 0 (mod 7) and 0 days after Friday is *Friday: 6 + 0 = 6*

4) 3,059
    ❖ Separate 3 from 059 and subtract: 59 - 3 = 56
    ❖ 56 ≡ 0 (mod 7) then 0 days after Saturday is *Saturday: 0 + 0 = 0*

5) 8,480
    ❖ Separate 8 from 480 and subtract: 480 − 8 = 472
    ❖ Separate 4 from 72. Doubling 4 yields 8. Add 8 + 72 = 80
    ❖ 80 ≡ 3 (mod 7) then 3 days after Thursday is *Sunday: 5 + 3 = 8 ≡ 1 (mod 7)*

6) 9,253

- ❖ Separate 9 from 253 and subtract: 253 – 9 = 244
- ❖ Separate 2 from 44. Doubling 2 yields 4. Add: 4 + 44 = 48
- ❖ $48 \equiv 6 \pmod 7$ then 6 days after Thursday is *Wednesday:*

5 + 6 = 11 ≡ 4 (mod 7)

7) 5,902

- ❖ Separate 5 from 902 and subtract: 902 – 5 = 897
- ❖ Separate 8 from 97. Doubling 8 yields 16. Add 16 + 97 = 113
- ❖ Separate 1 from 13. Doubling 1 yields 2. Add 2 + 13 = 15 ≡ 1 *(mod 7)*

Thus 1 day after Tuesday is Wednesday. 3 + 1 = 4.

8) 6,006

Separate 6 from 006 and subtract: 6 - 6 = 0 and we are done.

0 *days after Wednesday is Wednesday.*

9) 5,005

Separate 5 from 005 and subtract: 5 - 5 = 0 and we are *done.*

0 days after Tuesday is *Tuesday.*

10) 4,005

Separate 4 from 005 and subtract: 5 - 4 = 1 ≡ 1 (mod 7).

1 day after Thursday is Friday. Did you know the answer *immediately* upon *seeing* 4,005?

Final examination for Test A:

Find the day of the week for each date shown

1) December 7, 1941 (Pearl Harbor Day)
2) August 4, 1961 (President Obama's birthday)
3) July 4, 1776 (The ORIGINAL independence day)
4) April 20, 1930
5) February 28, 1956
6) August 10, 1945 (dropping of the A-bomb on Hiroshima)
7) November 24, 1981
8) June 13, 1858
9) June 13, 1958
10) September 11, 2001 (9/11)
11) January 1, 2002
12) October 30, 1999
13) September 18, 2019
14) February 29, 2004
15) May 14, 1962
16) November 22, 1963 (the assassination of John F. Kennedy)
17) July 20, 1969 (the moon landing)
18) August 19, 2005
19) April 22, 1975
20) February 28, 1900
21) June 21, 1982
22) October 3, 1945
23) January 23, 1976
24) March 19, 2000
25) September 9, 1981

Answers to TEST A:

Test A: Find the day of the week for each date shown.
The year codes will be *calculated*.

1) December 7, 1941 (Pearl Harbor Day)
   Calculating the code for 1941 using the year code algorithm:
   - ➢ $41 \div 4 = 10 \equiv 3 \pmod 7$
   - ➢ $41 \equiv 6 \pmod 7$
   - ➢ The 20th century code is 5.
   - ➢ $3 + 6 + 5 = 14 \equiv 0 \pmod 7$ so the code for 1941 is 0.

   Now we use the date algorithm.
   - ▪ The code for December is 1
   - ▪ The date is $7 \equiv 0 \pmod 7$
   - ▪ The year code is 0.
   - ▪ $1 + 7 + 0 \equiv 8 \equiv 1 \pmod 7$ thus, the answer is *Sunday*.

2) August 4, 1961 (President Obama's birthday)
   Calculating the code for 1961 using the year code algorithm:
   - ➢ $61 \div 4 = 15 \equiv 1 \pmod 7$
   - ➢ $61 \equiv 5 \pmod 7$
   - ➢ The 20th century code is 5
   - ➢ $1 + 5 + 5 = 11 \equiv 4 \pmod 7$ so the code for 1961 is 4.

   Now we use the date algorithm.
   - ▪ The code for August is 5.
   - ▪ The date is $4 \equiv 4 \pmod 7$
   - ▪ The year code is 4.
   - ▪ $5 + 4 + 4 \equiv 13 \equiv 6 \pmod 7$ Thus, the answer is *Friday*.

3) July 4, 1776 (The *ORIGINAL* independence day)
   Calculating the code for 1776 using the year code algorithm:
   ➤ $76 \div 4 = 19 \equiv 5 \pmod 7$
   ➤ $76 \equiv 6 \pmod 7$
   ➤ The 18th century code is 2.
   ➤ $5 + 6 + 2 = 13 \equiv 6 \pmod 7$ so *on July 4, the second code for 1776 is 6.*
   We don't *need* the first code, which is 5, because this is *not January or February.*
   Now we use the date algorithm.
   ▪ The code for July is 2.
   ▪ The date is $4 \equiv 4 \pmod 7$.
   ▪ The year code is 6.
   ▪ $2 + 4 + 6 \equiv 12 \equiv 5 \pmod 7$ thus, the answer is *Thursday.*

4) April 20, 1930
   Calculating the code for 1930 using the year code algorithm:
   ➤ $30 \div 4 = 7 \equiv 0 \pmod 7$
   ➤ $30 \equiv 2 \pmod 7$
   ➤ The 20th century code is 5
   ➤ $0 + 2 + 5 = 7 \equiv 0 \pmod 7$ so the code for 1930 is 0.
   Now we use the date algorithm.
   ▪ The code for April is 2.
   ▪ The date is $20 \equiv 6 \pmod 7$.
   ▪ The year code is 0.
   ▪ $2 + 6 + 0 \equiv 8 \equiv 1 \pmod 7$ thus, the answer is *Sunday.*

5) February 28, 1956
   Calculating the dual codes for 1956 using the year code algorithm:
   1956 has *two codes*. We will calculate the second code.
   ➤ $56 \div 4 = 14 \equiv 0 \pmod 7$
   ➤ $56 \equiv 0 \pmod 7$
   ➤ The 20th century code is 5.
   ➤ $0 + 0 + 5 \equiv 5 \pmod 7$ so the *second code* for 1956 is 5. The *first code* is 4.
     We must use the <u>*first*</u> code of 4 for February 28.
   Now we use the date algorithm.
   ▪ The code for February is 6.
   ▪ The date is $28 \equiv 0 \pmod 7$.
   ▪ The year code we are using is 4.
   ▪ $6 + 0 + 4 \equiv 10 \equiv 3 \pmod 7$ thus, the answer is *Tuesday.*

6) August 10, 1945 (dropping of the A-bomb on Hiroshima)

Calculating the year code for 1945:

➢ 1945: $45 \div 4 = 11 \equiv 4$ (mod 7)

➢ $45 \equiv 3$ (mod 7)

➢ The 20th century code is 5.

➢ $4 + 3 + 5 = 12 \equiv 5$ (mod 7) so the code for 1945 is 5.

Now we use the date algorithm.

▪ The code for August is 5.

▪ The date is $10 \equiv 3$ (mod 7).

▪ The year code is 5.

▪ $5 + 3 + 5 = 13 \equiv 6$ (mod 7) thus, the answer is *Friday*.

7) November 24, 1981

Calculating the year code for 1981:

➢ $81 \div 4 = 20 \equiv 6$ (mod 7)

➢ $81 \equiv 4$ (mod 7)

➢ The 20th century code is 5.

➢ $6 + 4 + 5 = 15 \equiv 1$ (mod 7) so the code for 1981 is 1.

Now we use the date algorithm.

▪ The code for November is 6.

▪ The date is $24 \equiv 3$ (mod 7).

▪ The year code is 1.

▪ $6 + 3 + 1 \equiv 10 \equiv 3$ (mod 7) thus, the answer is *Tuesday*.

8) June 13, 1858

Calculating the year code for 1858:

➢ $58 \div 4 = 14 \equiv 0$ (mod 7)

➢ $58 \equiv 2$ (mod 7)

➢ The 19th century code is 0.

➢ $0 + 2 + 0 \equiv 2$ (mod 7) so the code for 1858 is 2.

Now we use the date algorithm.

▪ The code for June is 0.

▪ The date is $13 \equiv 6$ (mod 7).

▪ The year code is 2.

▪ $0 + 6 + 2 = 8 \equiv 1$ (mod 7) thus, the answer is *Sunday*.

9) June 13, 1958

Calculating the year code for 1958:

➤ 58 ÷ 4 = 14 ≡ 0 (mod 7)

➤ 58 ≡ 2 (mod 7)

➤ The 20th century code is 5.

➤ 0 + 2 + 5 = 7 ≡ 0 (mod 7) so the code for 1958 is 0.

Now we use the date algorithm.

▪ The code for June is 0.

▪ The date is 13 ≡ 6 (mod 7).

▪ The year code is 0.

▪ 0 + 6 + 0 ≡ 6 (mod 7) thus, the answer is *Friday*.

Did you immediately know the answer from the answer to problem 8?

10) September 11, 2001 (9/11)

Calculating the year code for 2001:

➤ 01 ÷ 4 = 0 ≡ 0 (mod 7)

➤ 01 ≡ 1 (mod 7)

➤ The 21st century code is 4.

➤ 0 + 1 + 4 = 5 ≡ 5 (mod 7) so the code for 2001 is 5.

Now we use the date algorithm.

▪ The code for September is 1.

▪ The date is 11 ≡ 4 (mod 7).

▪ The year code is 5.

▪ 1 + 4 + 5 = 10 ≡ 3 (mod 7) thus, the answer is *Tuesday*.

11) January 1, 2002

Calculating the year code for 2002:

➤ 02 ÷ 4 = 0

➤ 02 ≡ 2 (mod 7)

➤ The 21st century code is 4.

➤ 0 + 2 + 4 ≡ 6 (mod 7) so the code for 2002 is 6.

Now we use the date algorithm.

▪ The code for January is 3.

▪ The date is 1 ≡ 1 (mod 7).

▪ 3 + 1 + 6 = 10 ≡ 3 (mod 7) thus, the answer is *Tuesday*.

12) October 30, 1999

Calculating the year code for 1999:

➤ $99 \div 4 = 24 \equiv 3 \pmod 7$
➤ $99 \equiv 1 \pmod 7$
➤ The 20th century code is 5.
➤ $3 + 1 + 5 = 9 \equiv 2 \pmod 7$ so the code for 1999 is 2.

Now we use the date algorithm.

■ The code for October is 3.
■ The date is $30 \equiv 2 \pmod 7$.
■ The year code is 2.
■ $3 + 2 + 2 = 7 \equiv 0 \pmod 7$ thus, the answer is *Saturday*.

13) September 18, 2019

Calculating the year code for 2019:

➤ $19 \div 4 = 4$.
➤ $19 \equiv 5 \pmod 7$.
➤ The 21st century code is 4.
➤ $4 + 5 + 4 = 13 \equiv 6 \pmod 7$ so the code for 2019 is 6.

Now we use the date algorithm.

■ The code for September is 1.
■ The date is $18 \equiv 4 \pmod 7$.
■ The year code is 6.
■ $1 + 4 + 6 = 11 \equiv 4 \pmod 7$ thus, the answer is *Wednesday*.

14) February 29, 2004

Calculating the year codes for 2004. This is a leap year.

➤ $04 \div 4 = 1$.
➤ $04 \equiv 4 \pmod 7$.
➤ The 21st century code is 4.
➤ $1 + 4 + 4 = 9 \equiv 2 \pmod 7$ so the *second code* for 2004 is 2.

Then the *first* code is 1. We must use *the first* code of 1 for February 29.

Now we use the date algorithm.

■ The code for February is 6.
■ The date is $29 \equiv 1 \pmod 7$.
■ The year code we are using is 1.
■ $6 + 1 + 1 \equiv 8 \equiv 1 \pmod 7$ thus, the answer is *Sunday*.

This problem was more difficult than the others.

15) May 14, 1962

Calculating the year code for 1962:

- ➢ $62 \div 4 = 15 \equiv 1$ (mod 7)
- ➢ $62 \equiv 6$ (mod 7)
- ➢ The 20th century code is 5
- ➢ $1 + 6 + 5 = 12 \equiv 5$ (mod 7) so the code for 1962 is 5.

Now we use the date algorithm.

- ▪ The code for May is 5.
- ▪ The date is $14 \equiv 0$ (mod 7).
- ▪ The year code is 5.
- ▪ $4 + 0 + 5 \equiv 9 \equiv 2$ (mod 7) thus, the answer is *Monday*.

16) November 22, 1963 (the assassination of President John F. Kennedy)

Calculating the year code for 1963:

- ➢ $63 \div 4 = 15 \equiv 1$ (mod 7).
- ➢ $63 \equiv 0$ (mod 7)
- ➢ The 20th century code is 5.
- ➢ $1 + 0 + 5 \equiv 6$ (mod 7) so the code for 1963 is 6.

Now we use the date algorithm.

- ▪ The code for November is 6.
- ▪ The date is $22 \equiv 1$ (mod 7).
- ▪ The year code is 6.
- ▪ $6 + 1 + 6 = 13 \equiv 6$ (mod 7) thus, the answer is *Friday*.

17) July 20, 1969 (the moon landing)

Calculating the year code for 1969:

- ➢ $69 \div 4 = 17 \equiv 3$ (mod 7)
- ➢ $69 \equiv 6$ (mod 7)
- ➢ The 20th century code is 5.
- ➢ $3 + 6 + 5 = 14 \equiv 0$ (mod 7) so the code for 1969 is 0.

Now we use the date algorithm.

- ▪ The code for July is 2.
- ▪ The date is $20 \equiv 6$ (mod 7).
- ▪ The year code is 0.
- ▪ $2 + 6 + 0 = 8 \equiv 1$ (mod 7) thus, the answer is *Sunday*.

18) August 19, 2005

Calculating the year cod for 2005:

- ➤ $05 \div 4 \equiv 1$ (mod 7).
- ➤ $05 \equiv 5$ (mod 7).
- ➤ The 21st century code is 4.
- ➤ $1 + 5 + 4 = 10 \equiv 3$ (mod 7) so the code for 2005 is 3.

Now we use the date algorithm.

- ▪ The code for August is 5.
- ▪ The date is $19 \equiv 5$ (mod 7).
- ▪ The year code is 3.
- ▪ $5 + 5 + 3 = 13 \equiv 6$ (mod 7) thus, the answer is *Friday*.

19) April 22, 1975

Calculating the year code for 1975:

- ➤ $75 \div 4 = 18 \equiv 4$ (mod 7).
- ➤ $75 \equiv 5$ (mod 7).
- ➤ The 20th century code is 5.
- ➤ $4 + 5 + 5 = 14 \equiv 0$ (mod 7) so the code for 1975 is 0.

Now we use the date algorithm.

- ▪ The code for April is 2.
- ▪ The date is $22 \equiv 1$ (mod 7).
- ▪ The year code is 0.
- ▪ $2 + 1 + 0 \equiv 3$ (mod 7) thus, the answer is Tuesday.

20) February 28, 1900

Calculating the year code for 1900:

- ➤ $00 \div 4 \equiv 0$ (mod 7).
- ➤ $00 \equiv 0$ (mod 7).
- ➤ The 20th century code is 5.

Note: We treat 1900 as the start of the 20th century even though it is *not*.

- ➤ $0 + 0 + 5 \equiv 5$ (mod 7) so the code for 1900 is 5.

   1900 was NOT a LEAP YEAR!

Now we use the date algorithm.

- ▪ The code for February is 6.
- ▪ The date is $28 \equiv 1$ (mod 7).
- ▪ The year code is 5.
- ▪ $6 + 0 + 5 = 11 \equiv 4$ (mod 7) thus, the answer is Wednesday.

21) June 21, 1982

Calculating the year code for 1982:

➢ $82 \div 4 = 20 \equiv 6$ (mod 7)

➢ $82 \equiv 5$ (mod 7)

➢ The 20th century code is 5.

➢ $6 + 5 + 5 = 16 \equiv 2$ (mod 7) so the code for 1982 is 2.

Now we use the date algorithm.

▪ The code for June is 0.

▪ The date is $21 \equiv 0$ (mod 7).

▪ The year code is 2.

▪ $0 + 0 + 2 \equiv 2$ (mod 7) thus, the answer is Monday.

22) October 3, 1945

Calculating the year code for 1945:

➢ $45 \div 4 = 11 \equiv 4$ (mod 7)

➢ $45 \equiv 3$ (mod 7).

➢ The 20th century code is 5.

➢ $4 + 3 + 5 = 12 \equiv 5$ (mod 7) so the code for 1945 is 5.

Now we use the date algorithm.

▪ The code for October is 3.

▪ The date is $3 \equiv 3$ (mod 7).

▪ The year code is 5.

▪ $3 + 3 + 5 = 11 \equiv 4$ (mod 7) thus, the answer is Wednesday.

23) January 23, 1976

Calculating the year codes for 1976. It is a leap year. We will calculate the *second code*.

➢ $76 \div 4 = 19 \equiv 5$ (mod 7).

➢ $76 \equiv 6$ (mod 7).

➢ The 20th century code is 5.

➢ $5 + 6 + 5 \equiv 16 \equiv 2$ (mod 7)

Thus, the *second* code for 1976 *from March to December is 2*. The first code is 1. We must use *the first* code of 1 for January.

Now we use the date algorithm.

▪ The code for January is 3.

▪ The date is $23 \equiv 2$ (mod 7).

▪ The year code is 1.

▪ $3 + 2 + 1 \equiv 6$ (mod 7) thus, the answer is *Friday*.

This is a very difficult problem. If you got it right, give yourself a pat on the back.

24) March 19, 2000

Calculate the codes for 2000: this is a leap year

➢ $00 \div 4 \equiv 0 \pmod 7$

➢ $00 \equiv 0 \pmod 7$

➢   The 21st century code is 4.

➢ $0 + 0 + 4 = 4 \equiv 4 \pmod 7$ so the *second* code for 2000 is 4. The *first code* is 3.

We must use the *second code* for March 19.

Now we use the date algorithm.

- The code for March is 6.
- The date is $19 \equiv 5 \pmod 7$.
- The year code is 4.
- $6 + 5 + 4 \equiv 15 \equiv 1 \pmod 7$ thus, the answer is Sunday.

25) September 9, 1981

Calculating the code for 1981:

➢ $81 \div 4 = 20 \equiv 6 \pmod 7$

➢ $81 \equiv 4 \pmod 7$

➢   The 20th century code is 5.

➢ $6 + 4 + 5 = 15 \equiv 1 \pmod 7$ so the code for 1981 is 1.

Now we use the date algorithm.

- The code for September is 1.
- The date is $9 \equiv 2 \pmod 7$.
- The year code is 1.
- $1 + 2 + 1 \equiv 4 \pmod 7$ thus, the answer is Wednesday.

Final examination for Test B:
The Appointment Calendar Problem

Find all dates in the given month and given year that fall on the given day of the week.

1) April 2005 Tuesday
2) February 1992 Wednesday
3) May 2011 Saturday
4) October 2000 Friday
5) August 2014 Thursday
6) February 2012 Monday
7) June 1943 Saturday
8) September 1975 Friday
9) November 2003 Wednesday
10) December 2008 Tuesday
11) January 1996 Saturday
12) November 2007 Thursday
13) July 2010 Wednesday
14) May 1951 Friday
15) August 2005 Wednesday
16) September 2001 Friday
17) February 1987 Thursday
18) Election Day is on the first Tuesday of November. On what date is Election Day in 2012?
19) Thanksgiving Day falls on the last Thursday of November. On what date will Thanksgiving Day be in 2011?
20) Daylight Saving time begins on the second Sunday of March. On what date will Daylight Saving begin in 2013?

Answers to Test B:
The Appointment Calendar Problem

Find all <u>dates</u> in the given month and given year that fall on the given day of the week.

1) April 2005 Tuesday
   - The code for April is 2.
   - The date is unknown, so we will call it x.
   - The code for 2005 is 3.
   - The code for Tuesday is 3.

   $2 + x + 3 \equiv 3 \pmod 7 \rightarrow x + 5 \equiv 3 \pmod 7 \rightarrow x = -2 \equiv 5 \pmod 7$

   Then *April 5, 12, 19 and 26* fall on Tuesday in 2005.

2) February 1992 Wednesday
   - The code for February is 6.
   - The date is x.
   - 1992 has *two codes: 0 and 1*. We *must* use the *first code* for February.
   - The code for Wednesday is 4.

   $6 + x + 0 \equiv 4 \pmod 7 \rightarrow x + 6 \equiv 4 \pmod 7 \rightarrow x = -2 \equiv 5 \pmod 7$.

   Thus *February 5, 12, 19 and 26* fall on Wednesday in 1992.

3) May 2011 Saturday
   - The code for May is 4.
   - The date is x.
   - The code for 2011 is 3.
   - The code for Saturday is 0.

   $4 + x + 3 \equiv 0 \pmod 7 \rightarrow x + 7 \equiv 0 \pmod 7 \rightarrow x + 0 \equiv 0 \pmod 7 \rightarrow x \equiv 0 \pmod 7$

   There is no 0th of May, so we will start with the 7th.

   Thus *May 7, 14, 21 and 28* fall on Saturday in 2011.

4) October 2000 Friday
    - The code for October is 3.
    - The date is x.
    - 2000 has *two codes: 3 and 4*. We will use 4 because this is October.
    - The code for Friday is 6.

$3 + x + 4 \equiv 6 \pmod 7 \rightarrow x + 0 \equiv 6 \pmod 7 \rightarrow x \equiv 6 \pmod 7$

Thus *October 6, 13, 20 and 27* fall on Friday in 2000.

5) August 2014 Thursday
    - The code for August is 5.
    - The date is x.
    - The code for 2014 is 0.
    - The code for Thursday is 5.

$5 + x + 0 \equiv 5 \pmod 7 \rightarrow x + 5 \equiv 5 \pmod 7 \rightarrow x \equiv 0 \pmod 7$

There is no 0th of August, so we will start with the 7th.

Thus *August 7, 14, 21 and 28* fall on Thursday in 2014

6) February 2012 Monday
    - The code for February is 6.
    - The date is x.
    - 2012 has *two codes: 4 and 5*.
      We must *use 4 because the date precedes March*.
    - The code for Monday is 2.

$6 + x + 4 \equiv 2 \pmod 7 \rightarrow x + 10 \equiv 2 \pmod 7 \rightarrow x \equiv -8 \pmod 7 \rightarrow x \equiv -1 \pmod 7$

Thus, $x \equiv 6 \pmod 7$. Thus *February 6, 13, 20 and 27* fall on Monday in 2012.

7) June 1943 Saturday
    - The code for June is 0.
    - The date is x.
    - The code for 1943 is 2.
    - The code for Saturday is 0.

$0 + x + 2 \equiv 0 \pmod 7 \rightarrow x + 2 \equiv 0 \pmod 7 \rightarrow x = -2 \equiv 5 \pmod 7$

Thus *June 5, 12, 19 and 26* fall on Saturday in 1943

8) September 1975 Friday
   - The code for September is 1.
   - The date is x.
   - The code for 1975 is 0.
   - The code for Friday is 6.

   $1 + x + 0 \equiv 6 \pmod 7 \rightarrow x + 1 \equiv 6 \pmod 7 \rightarrow x \equiv 5 \pmod 7$

   Thus *September 5, 12, 19 and 26* fall on Friday in 1975.

9) November 2003 Wednesday
   - The code for November is 6.
   - The date is x.
   - The code for 2003 is 0.
   - The code for Wednesday is 4.

   $6 + x + 0 \equiv 4 \pmod 7 \rightarrow x + 6 \equiv 4 \pmod 7 \rightarrow x = -2 \equiv 5 \pmod 7.$

   Thus November 5, 12, 19 and 26 fall on Wednesday in 2003.

10) December 2008 Tuesday
   - The code for December is 1.
   - The date is x.
   - 2008 has *two codes: 6 and 0.*
   - We must use the *second code* for December.
   - The code for Tuesday is 3.

   $1 + x + 0 \equiv 3 \pmod 7 \rightarrow x + 1 \equiv 3 \pmod 7 \rightarrow x \equiv 2 \pmod 7$

   Thus *December 2, 9, 16 and 23* fall on Tuesday in 2008.

11) January 1996 Saturday
   - The code for January is 3.
   - The date is x.
   - 1996 has *two codes: 5 and 6.*

   We must use the first code 5 because the month is January.
   - The code for Saturday is 0.

   $3 + x + 5 \equiv 0 \pmod 7 \rightarrow x + 8 \equiv 0 \pmod 7 \rightarrow x + 1 \equiv 0 \pmod 7 \rightarrow x = -1 \equiv 6 \pmod 7$

   Thus *January 6, 13, 20 and 27* fall on Saturday in 1996.

12) November 2007 Thursday

- The code for November is 6.
- The date is x.
- The code for 2007 is 5.
- The code for Thursday is 5.

$6 + x + 5 \equiv 5 \pmod 7 \rightarrow x + 11 \equiv 5 \pmod 7 \rightarrow x + 4 \equiv 5 \pmod 7 \rightarrow x \equiv 1 \pmod 7$

Thus *November 1, 8, 15 and 22* fall on Thursday in 2007.

13) July 2010 Wednesday

- The code for July is 2.
- The date is x.
- The code for 2010 is 2.
- The code for Wednesday is 4.

$2 + x + 2 \equiv 4 \pmod 7 \rightarrow x + 4 \equiv 4 \pmod 7 \rightarrow x \equiv 0 \pmod 7$

There is no 0th of July, so we will start with the 7th.

Thus *July 7, 13, 20 and 27* fall on Wednesday in 2010.

14) May 1951 Friday

- The code for May is 4.
- The date is x.
- The code for 1951 is 5.
- The code for Friday is 6.

$4 + x + 5 \equiv 6 \pmod 7 \rightarrow x + 9 \equiv 6 \pmod 7 \rightarrow x + 2 \equiv 6 \rightarrow x \equiv 4 \pmod 7$

Thus May 4, 11, 18 and 25 fall on Friday in 1951.

15) August 2005 Wednesday

- The code for August is 5.
- The date is x.
- The code for 2005 is 3.
- The code for Wednesday is 4.

$5 + x + 3 \equiv 4 \pmod 7 \rightarrow x + 8 \equiv 4 \pmod 7 \rightarrow x + 1 \equiv 4 \pmod 7 \rightarrow x \equiv 3 \pmod 7$

Thus August 3, 10, 17 and 24 fall on Wednesday in 2005.

16) September 2001 Friday
- The code for September is 1.
- The date is x.
- The code for 2001 is 5.
- The code for Friday is 6.

$1 + x + 5 \equiv 6 \pmod 7 \rightarrow x + 6 \equiv 6 \pmod 7 \rightarrow x \equiv 0 \pmod 7$

There is no 0th of September, so we will start with the 7th.

Thus September 7, 14, 21 and 28 fall on Friday in 2001.

17) February 1987 Thursday
- The code for February is 6.
- The date is x.
- The code for 1987 is 1.
- The code for Thursday is 5.

$6 + x + 1 \equiv 5 \pmod 7 \rightarrow x + 7 \equiv 5 \pmod 7 \rightarrow x + 0 \equiv 5 \pmod 7 \rightarrow x \equiv 5 \pmod 7$

Thus *February 5, 12, 19 and 26 fall on Thursday in 1987.*

18) Election Day is on the first Tuesday of November.
- On what date is Election Day in 2012?
- The code for November is 6.
- The date is unknown so it is x.
- *The codes for 2012 are 4 and 5.*

We will use the *second code* for 2012 because November is after February.
- The code for Tuesday is 3.

$6 + x + 5 \equiv 3 \pmod 7 \rightarrow x + 11 \equiv 3 \pmod 7 \rightarrow x + 4 \equiv 3 \pmod 7 \rightarrow x = -1 \equiv 6 \pmod 7$

Thus, Tuesday, *November 6, 2012* will be Election Day.

19) Thanksgiving Day falls on the last Thursday of November.
On what date will Thanksgiving Day be in 2011?
- The code for November is 6.
- The date is x.
- The code for 2011 is 3.
- The code for Thursday is 5.

$6 + x + 3 \equiv 5 \pmod 7 \rightarrow x + 9 \equiv 5 \pmod 7 \rightarrow x + 2 \equiv 5 \pmod 7 \rightarrow x \equiv 3 \pmod 7$

Thus, November 3, 2011 falls on a Thursday. Add 21 days to reach Thanksgiving Day.

$3 + 21 = 24$ thus, *Thanksgiving Day is Thursday, November 24, 2011.*

20) Daylight Saving time begins on the second Sunday of March.

On what date will Daylight Saving begin in 2013?

- The code for March is 6.
- The date is x.
- The code for 2013 is 6.
- The code for Sunday is 1.

$6+x+6\equiv1\,(\mathrm{mod}\,7)\rightarrow x+12\equiv1\,(\mathrm{mod}\,7)\rightarrow x+5\equiv1\,(\mathrm{mod}\,7)\rightarrow x=-4\equiv3\,(\mathrm{mod}\,7)$

Thus, March 3, 2013 is the *FIRST Sunday in March in 2013*.

The <u>*second*</u> Sunday of *March, March 10, 2013 starts Daylight Saving time.*

Final examination for Test C:
The Law & Order Problem

Use the years from 2000 to 2009 to find a year or years in which the following dates were possible. There are usually two, but for some, there is only one.

1) Thursday, April 10

2) Tuesday, April 10

3) Friday, May 12

4) Saturday, September 19

5) Wednesday, February 20

6) Thursday, August 16

7) Sunday, March 30

8) Tuesday, December 22

9) Monday, July 14

10) Saturday, April 6

11) Sunday, January 12

12) Friday, June 4

13) Friday, November 28

14) Wednesday, March 31

15) Tuesday, April 29

16) Saturday, August 4

17) Thursday, June 18

18) Friday, October 24

19) Wednesday, March 2

20) Saturday, December 31

Answers to Test C:
The Law & Order Problem

Using the years 2000 to 2009, inclusive, find all years on which the date could have occurred.

In each example, the code of the year is represented by x.

1) Thursday, April 10:
   - The code for April is 2.
   - The date is $10 \equiv 3 \pmod 7$.
   - The year code is unknown, so it is x.
   - The code for Thursday is 5.

   $2 + 3 + x \equiv 5 \rightarrow x + 5 \equiv 5 \pmod 7 \rightarrow x \equiv 0 \pmod 7$

   The year code must be 0. *2003 has the code 0.*

   Also from March to December, the *second code* for 2008 is 0.

   The years 2003 and 2008 are the solutions.

2) Tuesday, April 10:
   - The code for April is 2.
   - The year code is x.
   - The date is $10 \equiv 3 \pmod 7$.
   - The code for Tuesday is 3.

   $2 + 3 + x \equiv 3 \pmod 7 \rightarrow x + 5 \equiv 3 \pmod 7 \rightarrow x = -2 \equiv 5 \pmod 7$

   Thus, the year code must be 5. 2001 and 2007 have the code 5.

   2001 and 2007 are solutions.

3) Friday, May 12:
   - The code for May is 4.
   - The date is $12 \equiv 5 \pmod 7$.
   - The year code is x.
   - The code for Friday is 6.

   $4 + 5 + x \equiv 6 \pmod 7 \to x + 9 \equiv 6 \pmod 7 \to x + 2 \equiv 6 \pmod 7 \to x \equiv 4 \pmod 7$

   The year code must be 4. *The second code* of 2000 is 4 and 2006 has the code 4.

   The years 2000 and 2006 are solutions.

4) Saturday, September 19:
   - The code for September is 1.
   - The date is $19 \equiv 5 \pmod 7$.
   - The year code is x.
   - The code for Saturday is 0.

   $1 + 5 + x \equiv 0 \pmod 7 \to x + 6 \equiv 0 \pmod 7 \to x = -6 \equiv 1 \pmod 7$

   The year code must be 1. *2009 is the only* solution.

5) Wednesday, February 20:
   - The code for February is 6.
   - The date is $20 \equiv 6 \pmod 7$.
   - The year code is x.
   - The code for Wednesday is 4.

   $6 + 6 + x \equiv 4 \pmod 7 \to x + 12 \equiv 4 \pmod 7 \to x + 5 \equiv 4 \pmod 7 \to x = -1 \equiv 6 \pmod 7$

   The year code must be 6. The code of 2002 is 6. The *first code* of 2008 *also* works.

   2002 and 2008 are solutions. Kudos to those who got this right.

6) Thursday, August 16:
   - The code for August is 5.
   - The date is $16 \equiv 2 \pmod 7$.
   - The year code is x.
   - The code for Thursday is 5.

   $5 + 2 + x \equiv 5 \pmod 7 \to x + 7 \equiv 5 \pmod 7 \to x + 0 \equiv 5 \pmod 7 \to x \equiv 5 \pmod 7$

   The year code must be 5. The codes of 2001 and 2007 are 5.

   2001 and 2007 are the solutions.

7) Sunday, March 30:
-   The code for March is 6.
-   The date is $30 \equiv 2 \pmod 7$.
-   The year code is x.
-   The code for Sunday is 1.

$6 + 2 + x \equiv 1 \pmod 7 \rightarrow x + 8 \equiv 1 \pmod 7 \rightarrow x + 1 \equiv 1 \pmod 7 \rightarrow x \equiv 0 \pmod 7$

The year code must be 0.

The code of 2003 is 0. The *second code* for 2008 works for March.

2003 and 2008 are the solutions.

8) Tuesday, December 22:
-   The code for December is 1.
-   The date is $22 \equiv 1 \pmod 7$.
-   The year code is x.
-   The code for Tuesday is 3.

$1 + 1 + x \equiv 3 \pmod 7 \rightarrow x + 2 \equiv 3 \pmod 7 \rightarrow x \equiv 1 \pmod 7$

The year code must be 1. 2009 is the *only* solution.

9) Monday, July 14:
-   The code for July is 2.
-   The date is $14 \equiv 0 \pmod 7$.
-   The year code is x.
-   The code for Monday is 2.

$2 + 0 + x \equiv 2 \pmod 7 \rightarrow x + 2 \equiv 2 \pmod 7 \rightarrow x \equiv 0 \pmod 7$

The year code must be 0. 2003 and the *second part of* 2008 are solutions.

10) Saturday, April 6:
-   The code for April is 2.
-   The date is 6.
-   The year code is x.
-   The code for Saturday is 0.

$2 + 6 + x \equiv 0 \pmod 7 \rightarrow x + 8 \equiv 0 \pmod 7 \rightarrow x + 1 \equiv 0 \pmod 7 \rightarrow x = -1 \equiv 6 \pmod 7$

The year code must be 6 thus 2002 is the *only* solution.

11) Sunday, January 12:
- The code for January is 3.
- The date is $12 \equiv 5 \pmod 7$.
- The year code is x.
- The code for Sunday is 1.

$3 + 5 + x \equiv 1 \pmod 7 \rightarrow x + 8 \equiv 1 \pmod 7 \rightarrow x + 1 \equiv 1 \pmod 7 \rightarrow x \equiv 0 \pmod 7$

The year code must be 0 thus 2003 is the *only* solution.

Up to 2009, no leap year this century has a *first code* of 0.

12) Friday, June 4:
- The code for June is 0.
- The date is 4.
- The year code is x.
- The code of Friday is 6.

$0 + 4 + x \equiv 6 \pmod 7 \rightarrow x + 4 \equiv 6 \pmod 7 \rightarrow x \equiv 2 \pmod 7$

The year code must be 2. The *second* code of 2004 is 2.

2004 is the solution.

13) Friday, November 28:
- The code for November is 6.
- The date is $28 \equiv 0 \pmod 7$.
- The year code is x.
- The code for Friday is 6

$6 + 0 + x \equiv 6 \pmod 7 \rightarrow x + 6 \equiv 6 \pmod 7 \rightarrow x \equiv 0 \pmod 7$

The year code must be 0. Thus, 2003 is a solution.

The *second* code of 2008 is 0.

2003 and 2008 are solutions.

14) Wednesday, March 31:
- The code for March is 6.
- The date is $31 \equiv 3 \pmod 7$.
- The year code is x.
- The code for Wednesday is 4.

$6 + 3 + x \equiv 4 \pmod 7 \rightarrow x + 9 \equiv 4 \pmod 7 \rightarrow x + 2 \equiv 4 \pmod 7 \rightarrow x \equiv 2 \pmod 7$

The year code must be 2 thus 2004 is the *only* solution with a *second code* of 2.

15) Tuesday, April 29:
- The code for April is 2.
- The date is $29 \equiv 1 \pmod 7$.
- The year code is x.
- The code for Tuesday is 3.

$2 + 1 + x \equiv 3 \rightarrow x + 3 \equiv 3 \pmod 7 \rightarrow x \equiv 0 \pmod 7$

The year code must be 0, which is the code of 2003.

The *second code* of 2008 is 0.

2003 and 2008 are the solutions.

16) Saturday, August 4:
- The code for August is 5.
- The date is 4.
- The year code is x.
- The code for Saturday is 0.

$5 + 4 + x \equiv 0 \pmod 7 \rightarrow x + 9 \equiv 0 \pmod 7 \rightarrow x + 2 \equiv 0 \pmod 7 \rightarrow x = -2 \equiv 5 \pmod 7$

The year code must be 5 thus, 2001 and 2007 are the solutions.

17) Thursday, June 18:
- The code for June is 0.
- The date is $18 \equiv 4 \pmod 7$.
- The year code is x.
- The code for Thursday is 5.

$0 + 4 + x \equiv 5 \rightarrow x + 4 \equiv 5 \pmod 7 \rightarrow x \equiv 1 \pmod 7$

The year code must be 1 thus 2009 is the *only* solution.

18) Friday, October 24:
- The code for October is 3.
- The date is $24 \equiv 3 \pmod 7$.
- The year code is x.
- The code for Friday is 6.

$3 + 3 + x \equiv 6 \pmod 7 \rightarrow x + 6 \equiv 6 \pmod 7 \rightarrow x \equiv 0 \pmod 7$

The year code must be 0, the code of 2003. The *second code* of 2008 is 0.

The solutions are 2003 and 2008.

19) Wednesday, March 2:

- The code for March is 6.
- The date is 2.
- The year code is x.
- The code for Wednesday is 4.

$6 + 2 + x \equiv 4 \rightarrow x + 8 \equiv 4 \pmod 7 \rightarrow x + 1 \equiv 4 \pmod 7 \rightarrow x \equiv 3 \pmod 7$

The year code must be 3 thus, 2005 is the *only* solution.

20) Saturday, December 31:

- The code for December is 1.
- The date is $31 \equiv 3 \pmod 7$.
- The year code is x.
- The code for Saturday is 0.

$1 + 3 + x \equiv 0 \pmod 7 \rightarrow x + 4 \equiv 0 \pmod 7 \rightarrow x = -4 \equiv 3 \pmod 7$

The year code must be 3 so once again, 2005 is the only solution

Final examination for Test D:
Finding future years to re-use the calendar

Find *two years* after the given year that will have the same calendar.

1) 1951

2) 2003

3) 1974

4) 1969

5) 1900

6) 2009

7) 1989

8) 2022

9) 1997

10) 1933

11) 1955

12) 1999

13) 1923

14) 2007

15) 1962

16) 1947

17) 1975

18) 1981

19) 1959

20) 1978

Answers to Test D:
Finding future years to RE-USE the calendar

Find TWO YEARS after the given year that will have the SAME CALENDAR.

1) 1951 is an LY+3 year which means that you must add 11.
   1951 + 11 = *1962 is identical to* 1951.
   1962 is an LY+2, so add 11 *again*. 1962 + 11 = *1973 and is identical to* 1962.
   Then 1951, 1962 and 1973 are all identical. Their common code is 5.

2) 2003 is an LY+3 year which means that you must add 11.
   2003 + 11 = *2014 is identical to* 2003.
   2014 is an LY+2, so add 11 *again*. 2014 + 11 = *2025 and is identical to* 2014.
   Then 2003, 2014 and 2025 are all identical. Their common code is 0.

3) 1974 is an LY+2 year which means that you must add 11.
   1974 + 11 = *1985 and is identical to* 1974.
   1985 is an LY+1, so add 6. 1985 + 6 = *1991 and is identical to* 1985.
   Then 1974, 1985 and 1991 are all identical. Their common code is 6.

4) 1969 is an LY+1 year which means that you must add 6.
   1969 + 6 = *1975 and is identical to* 1969.
   1975 is an LY+3, so add 11. 1975 + 11 = *1986 and is identical to* 1975.
   Then 1969, 1975 and 1986 are all identical. Their common code is 0.

5) *1900 is a special case.* It is not an LY+1 or an LY+2 or an LY+3 or a leap year.
   Adding 6 is required because it is in the <u>last decade </u>of the 19th century.
   1900 + 6 = *1906 and is identical to* 1900. 1906 is an LY+2. Relax.
   Now everything's normal. This is the 20th century. Add 11.
   1906 + 11 = *1917 and is identical to* 1906.
   Then 1900, 1906 and 1917 are all identical. Their common code is 5.

6) 2009 is an LY+1 year which means that you must add 6.
    2009 + 6 = 2015 and *is identical to* 2009.
    2015 is an LY+3, so add 11. 2015 + 11 = *2026 and is identical to* 2015.
    Then 2009, 2015 and 2026 are all identical. Their common code is 1.

7) 1989 is an LY+1 year which means that you must add 6.
    1989 + 6 = *1995 and is identical to* 1989.
    1995 is an LY+3, so add 11. 1995 + 11 = *2006 and is identical to* 1995.
    Then 1989, 1995 and 2006 are all identical. Their common code is 4.

8) 2022 is an LY+2 year which means that you must add 11.
    2022 + 11 = *2033 and is identical to* 2022.
    2033 is an LY+1, so add 6. 2033 + 6 = *2039 and is identical to* 2033.
    Then 2022, 2033 and 2039 are all identical. Their common code is 3.

9) 1997 is an LY+1 year which means that you must add 6.
    1997 + 6 = *2003 and is identical to* 1997.
    2003 is an LY+3, so add 11. 2003 + 11 = *2014 and is identical to* 2003.
    Then 1997, 2003 and 2014 are all identical. Their common code is 0.

10) 1933 is an LY+1 year which means that you must add 6.
    1933 + 6 = *1939 and is identical to* 1933.
    1939 is an LY+3, so add 11. 1939 + 11 = *1950 and is identical to* 1939.
    Then 1933, 1939 and 1950 are all identical. Their common code is 4.

11) 1955 is an LY+3 year which means that you must add 11.
    1955 + 11 = *1966 and is identical to* 1955.
    1966 is an LY+2, so add 11. 1966 + 11 = *1977 and is identical to* 1966.
    Then 1955, 1966 and 1977 are all identical. Their common code is 3.

12) 1999 is an LY+3 year which means that you must add 11.
    1999 + 11 = *2010 and is identical to* 1999.
    2010 is an LY+2, so add 11. 2010 + 11 = *2021 and is identical to* 2010.
    Then 1999, 2010 and 2021 are all identical. Their common code is 2.

13) 1923 is an LY+3 year which means that you must add 11.

1923 + 11 = *1934 and is identical to* 1923.

1934 is an LY+2, so add 11. 1934 + 11 = *1945 and is identical to* 1934.

Then 1923, 1934 and 1945 are all identical. Their common code is 5.

14) 2007 is an LY+3 year which means that you must add 11.

2007 + 11 = *2018 and is identical to* 2007.

2018 is an LY+2, so add 11. 2018 + 11 = *2029 and is identical to* 2018.

Then 2007, 2018 and 2029 are all identical. Their common code is 5.

15) 1962 is an LY+2 year which means that you must add 11.

1962 + 11 = *1973 and is identical to* 1962.

1973 is an LY+1, so add 6. 1973 + 6 = *1979 and is identical to* 1973.

Then 1962, 1973 and 1979 are all identical. Their common code is 5.

16) 1947 is an LY+3 year which means that you must add 11.

1947 + 11 = *1958 and is identical to* 1947.

1958 is an LY+2, so add 11. 1958 + 11 = *1969 and is identical to* 1958.

Then 1947, 1958 and 1969 are all identical. Their common code is 0.

17) 1975 is an LY+3 year which means that you must add 11.

1975 + 11 = *1986 and is identical to* 1975.

1986 is an LY+2, so add 11. 1986 + 11 = *1997 and is identical to* 1986.

Then 1975, 1986 and 1997 are all identical. Their common code is 0.

18) 1981 is an LY+1 year which means that you must add 6.

1981 + 6 = *1987 and is identical to* 1981.

1987 is an LY+3, so add 11. 1987 + 11 = *1998 and is identical to* 1987.

Then 1981, 1987 and 1998 are all identical. Their common code is 1.

19) 1959 is an LY+3 year which means that you must add 11.

1959 + 11 = *1970 and is identical to* 1959.

1970 is an LY+2, so add 11. 1970 + 11 = *1981 and is identical to* 1970.

Then 1959, 1970 and 1981 are all identical. Their common code is 1.

20) 1978 is an LY+2 year which means that you must add 11.

1978 + 11 = *1989 and is identical to* 1978.

1989 is an LY+1, so add 6. 1989 + 6 = *1995 and is identical to* 1989.

Then 1978, 1989 and 1995 are all identical. Their common code is 4.

Review these answers for years and year codes that come up repeatedly.

Final examination: Test E.

Find the month(s), if any, in which a given date in a given year can fall on the given day of the week.

1) The 31st falls on a Tuesday in 1998

2) The 23rd falls on a Thursday in 1989

3) The 15th falls on a Sunday in 2005

4) The 26th falls on a Wednesday in 2010

5) The 25th falls on a Saturday in 2011

6) The 31st falls on a Thursday in 1999

7) The 13th falls on a Tuesday in 2010

8) The 5th falls on a Monday in 2004

9) The 30th falls on a Wednesday in 2002

10) The 19th falls on a Thursday in 2000

Test E: Answers

1) The 31st falls on a Tuesday in 1998.
   - The month code is x.
   - The date is $31 \equiv 3$ (mod 7).
   - The code for 1998 is 1.
   - The code for Tuesday is 3.
     $$x + 3 + 1 \equiv 3 \ (mod \ 7) \rightarrow x + 4 \equiv 3 \ (mod \ 7) \rightarrow x \equiv \text{-}1 \equiv 6 \ (mod \ 7)$$
     The only month with a code of 6 *that has 31 days* is *March*.

2) The 23rd falls on a Thursday in 1989.
   - The month code is x.
   - The date is $23 \equiv 2$ (mod 7).
   - The code for 1989 is 4.
   - The code for Thursday is 5.
   $$x + 2 + 4 \equiv 5 \ (mod \ 7) \rightarrow x + 6 \equiv 5 \ (mod \ 7) \rightarrow x \equiv \text{-}1 \equiv 6 \ (mod \ 7)$$
   February, March and November are the solutions.

3) The 15th falls on a Sunday in 2005.
   - The month code is x.
   - The date is $15 \equiv 1$ (mod 7).
   - The code for 2005 is 3.
   - The code for Sunday is 1.
   $$x + 1 + 3 \equiv 1 \ (mod \ 7) \rightarrow x + 4 \equiv 1 \ (mod \ 7) \rightarrow x \equiv \text{-}3 \equiv 4 \ (mod \ 7)$$
   May is the solution.

4) The 26th falls on a Wednesday in 2010.
   - The month code is x.
   - The date is $26 \equiv 5 \pmod 7$.
   - The code for 2010 is 2.
   - The code for Wednesday is 4.
     $x + 5 + 2 \equiv 4 \pmod 7 \rightarrow x + 0 \equiv 4 \pmod 7 \rightarrow x \equiv 4 \pmod 7$
     Again, May is the solution.

5) The 25th falls on a Saturday in 2011.
   - The month code is x.
   - The date is $25 \equiv 4 \pmod 7$.
   - The code for 2011 is 3.
   - The code for Saturday is 0.
     $x + 4 + 3 \equiv 0 \pmod 7 \rightarrow x + 0 \equiv 0 \pmod 7 \rightarrow x \equiv 0 \pmod 7$
     June is the solution.

6) The 31st falls on a Thursday in 1999.
   - The month code is x.
   - The date is $31 \equiv 3 \pmod 7$.
   - The code for 1999 is 2.
   - The code for Thursday is 5.
     $x + 3 + 2 \equiv 5 \pmod 7 \rightarrow x + 5 \equiv 5 \pmod 7 \rightarrow x \equiv 0 \pmod 7$
     The only month with the code of 0 is June.
     However, June is NOT a solution because *it only has 30 days!*
     There is no solution.

7) The 13th falls on a Tuesday in 2010.
   - The month code is x.
   - The date is $13 \equiv 6 \pmod 7$.
   - The code for 2010 is 2.
   - The code for Tuesday is 3.
     $x + 6 + 2 \equiv 3 \pmod 7 \rightarrow x + 8 \equiv 3 \pmod 7 \rightarrow x + 1 \equiv 3 \pmod 7 \rightarrow x \equiv 2 \pmod 7$
     April and July are solutions.

8) The 5th falls on a Monday in 2004.

- The month code is x.
- The date, 5, does not require reducing.
- The dual codes for 2004 are 1 and 2.
- The code for Monday is 2.
- Using the *first* code of 1:

  $x + 5 + 1 \equiv 2 \pmod 7 \rightarrow x + 6 \equiv 2 \pmod 7 \rightarrow x \equiv -4 \equiv 3 \pmod 7$

  January is a solution.
- Using the second code of 2:

  $x + 5 + 2 \equiv 2 \pmod 7 \rightarrow x + 0 \equiv 2 \pmod 7 \rightarrow x \equiv 2 \pmod 7$

  April and July are solutions.

In all, there are *three* solutions: January, April and July.

9) The 30th falls on a Wednesday in 2002.

- The month code is x.
- The date is $30 \equiv 2 \pmod 7$.
- The code for 2002 is 6.
- The code for Wednesday is 4.

  $x + 2 + 6 \equiv 4 \pmod 7 \rightarrow x + 8 \equiv 4 \pmod 7 \rightarrow x + 1 \equiv 4 \pmod 7 \rightarrow x \equiv 3 \pmod 7$

  January and October are solutions.

10) The 19th falls on a Thursday in 2000.

- The month code is x.
- The date is $19 \equiv 5 \pmod 7$.
- 2000 has the dual codes 3 and 4.
- The code for Thursday is 5.
- Using the first code:

  $x + 5 + 3 \equiv 5 \pmod 7 \rightarrow x + 8 \equiv 5 \pmod 7 \rightarrow x + 1 \equiv 5 \pmod 7 \rightarrow x \equiv 4 \pmod 7$

  Neither January nor February has the code of 4. There is no solution yet.
- Using the second code:

  $x + 5 + 4 \equiv 5 \pmod 7 \rightarrow x + 9 \equiv 5 \pmod 7 \rightarrow x + 2 \equiv 5 \pmod 7 \rightarrow x \equiv 3 \pmod 7$

  October is the *only solution*. January is the wrong part of the year.

Final examination: Test F:
Reducing LARGE NUMBERS in Modulo 7

Reduce in modulo 7

1) 825

2) 546

3) 427

4) 965

5) 634

6) 5,018

7) 9,107

8) 6,289

9) 12,754

10) 14,927

Answers to Test F: Reduce in modulo 7.
No shortcuts were taken. That is your job.

1) 825: separate 8 from 25. Doubling 8 yields 16, then add 16 + 25 = 41.
    Reduce 41 ≡ 6 (mod 7).

2) 546: separate 5 from 46. Doubling 5 yields 10, then add 10 + 46 = 56.
    Reduce 56 ≡ 0 (mod 7).

3) 427: separate 4 from 27. Doubling 4 yields 8, then add 8 + 27 = 35.
    Reduce 35 ≡ 0 (mod 7).

4) 965: separate 9 from 65. Doubling 9 yields 18, then add 18 + 65 = 83.
    Reduce 83 ≡ 6 (mod 7).

5) 634: separate 6 from 34. Doubling 6 yields 12, then add 12 + 34 = 46.
    Reduce 46 ≡ 4 (mod 7).

6) 5,018: separate 5 from 018. *Subtract* 18 - 5 = 13. Reduce 13 ≡ 6 (mod 7).

7) 9,107: separate 9 from 107. *Subtract* 107 - 9 = 98 ≡ 0 (mod 7).
    If you look again at 9,107, you will see 91 and 07, *both multiples of 7!*

8) 6,289: separate 6 from 289. *Subtract* 289 - 6 = 283.
    Now the problem is 283. Separate 2 from 83.
    Doubling 2 yields 4, then add 4 + 83 = 87.
    Reduce 87 ≡ 3 (mod 7).

9) 12,754: separate 12 from 754. *Subtract* 754 - 12 = 742.
Now the problem is 742. Separate 7 from 42.
Doubling 7 yields 14, then add 14 + 42 = 56.
Reduce 56 ≡ 0 (mod 7).
You might have known the answer immediately upon seeing 742.

10) 14,927: separate 14 from 927. *Subtract* 927 - 14 = 913.
Now the problem is 913: separate 9 from 13.
Doubling 9 yields 18, then add 18 + 13 = 31.
Reduce 31 ≡ 3 (mod 7).
You might have *scratched the 14* (in 14,927)) leaving 927.
Change 7 to 0 and 9 to 2.
Then the number is whittled down to 220.
Separate 2 from 20 and add 4 + 20 = 24 ≡ 3 (mod 7).

Cumulative examination

1) Halloween falls on October 31. On what day of the week does it fall in 2015?

2) The Supreme Court begins its term on the first Monday in October. On what date will the Supreme Court begin its term in 2018?

3) On what date is the last Saturday of 2013?

4) Sasha was born on July 11, 1980. Her brother, Homer, was born exactly 600 days after the day she was born. On what day of the week was Homer born?

5) My doctor sees me only on Mondays. I saw him on Monday, December 7, 2009. Upon leaving his office, he told me to make an appointment to see him in four months. In what month will I make that appointment? What will be the date?

6) You found a calendar for 1993 in your attic. In what year or years will it be reusable in the 21st century? Stop at whatever year is current.

7) From 1990 to 1999, in what years was Saturday the 31st impossible?

8) In what month(s) could Friday the 13th occur in 2020?

9) March 1, 1780 fell on a Wednesday. On what day of the week will March 1, 2080 fall?

10) Francine just celebrated her birthday. Happy birthday, Francine. She tells her friends that she is her early thirties but younger than 35. She was born on Monday, June 9.
Find the year in which she was born.

Answers to the cumulative examination

1) Halloween falls on October 31. On what day of the week does it fall in 2015?
   * The month code is 3.
   * The date is $31 \equiv 3 \pmod 7$.
   * The year code is 1
     $3 + 3 + 1 \equiv 7 \equiv 0 \pmod 7$ thus, the answer is Saturday whose code is 0.

2) The Supreme Court begins its term on the first Monday in October.
   On what date will the Supreme Court begin its term on in 2018?
   * The month code is 3.
   * The date is x.
   * The year code is 5.
   * The code for Monday is 2.
     $3 + x + 5 \equiv 2 \pmod 7 \rightarrow x + 8 \equiv 2 \pmod 7 \rightarrow x + 1 \equiv 2 \pmod 7$
     $x \equiv 1 \pmod 7$
     Thus, on October 1, 2018, the Supreme Court begins it term.

3) On what date is the last Saturday of 2013?
   * The month code is 1 (because December is the last month!).
   * The date is x, but it must one of these: Dec 25, 26, 27, 28, 29, 30 or 31.
   * The year code is 6
   * The code of the day of the week is 0.
   $1 + x + 6 \equiv 0 \pmod 7 \rightarrow x + 0 \equiv 0 \pmod 7 \rightarrow x \equiv 0 \pmod 7$
   There is no 0th date, so we begin with the 7th and add 7 until we reach the last Saturday.
   December 28, 2013 is the last Saturday in 2018.

4) Sasha was born on July 11, 1980. Her brother, Homer, was born exactly 600 days after the day she was born. On what day of the week was Homer born? First, find the day of the week on which Sasha was born.

❖ The month code is 2.

❖ The date is $11 \equiv 4 \pmod 7$.

❖ The year code is 0.

$2 + 4 + 0 \equiv 6 \pmod 7$ thus, Sasha was born on a Friday, whose code is 6. Now reduce 600 (mod 7). Separate 6 from 00. Double 6 and add the result, 12 to 00. You get 12 which quickly reduces to 5 (mod 7).

5 days after Friday is Wednesday.

Homer was born on a Wednesday.

5) My doctor sees me only on Mondays. I saw him on Monday, December 7, 2009. Upon leaving his office, he told me to make an appointment to see him in four months. In what month will I make that appointment? What will be the date?

The month is found by adding 4 to 12, the number of December is the sequence of the months. The sum is 16. You need *modulo 12*. Subtract 12 from 16 and you get 4. The month that is 4th in the sequence of months is April.

❖ The month *code* is 2.

❖ The date is x.

❖ The year code is 2 (for 2010).

❖ The code for Monday is 2.

$2 + x + 2 \equiv 2 \pmod 7 \rightarrow x + 4 \equiv 2 \pmod 7 \rightarrow x \equiv -2 \equiv 5 \pmod 7$

Then the appointment will be made for April 5, 12, 19 or 26 in 2010.

6) You found a calendar for 1993 in your attic. In what year or years will it be reusable in the early 21st century?

This will be solved one way only. Other methods are welcome.

1993 is an LY+1 year, so add 6. 1999 is the result. 1999 is an LY+3 year, so add 11. 2010 is the result and is the solution. You may find more.

7) From 1990 to 1999, in what year or years was Saturday the 31st *impossible*?
The code for Saturday is 0. The rule says to add 3 to the year code.
Solve $x + 3 = 0 \rightarrow x \equiv -3$ (mod 7) $\rightarrow x \equiv 4$ (mod 7).
Thus, the year must have the code of 4.
The starting year, 1990 has the code $4 + 1 = 5$.
Repeated addition of 1 starting with the code for 1990 is shown with the codes:
1990(5) then 1991(6) then 1992(0, 1) then 1993(2) then 1994(3) and 1995(4) Stop.
The year in which Saturday the 31st was impossible is 1995.

8) In what month(s) could Friday the 13th occur in 2020?
The *dual* year code for 2020 is (1, 2). The complement of 1 is 6 so February 13 falls on a Friday in 2020. The complement of 2 is 5, so August 13, 2020 falls on a Friday.

9) March 1, 1780 fell on a Wednesday.
On what day of the week will March 1, 2080 fall?
The day of the week advances 5 days from 1780 to 1880 and 5 days again to 1980 and then 6 days to 2080.
In all, the day of the week advances $5 + 5 + 6 = 16$ days.
Reducing, $16 \equiv 2$ (mod 7).
So March 1, 2080 advances 2 days from Monday. It falls on Wednesday.

10) Francine just celebrated her birthday in 2009. Happy birthday, Francine.
She tells her friends that she is her early thirties but younger than 35. She was born on Monday, June 9.
Find the year in which she was born.
Since $2009 - 35 = 1974$, we start there. The code for 1974 is 6 and the code for 1975 is 0.
  ❖ The month code is 0.
  ❖ The date is $9 \equiv 2$ (mod 7)
  ❖ The year code is x.
  ❖ The code for Monday is 2.
    $0 + 2 + x \equiv 2$ (mod 7) $\rightarrow x + 2 \equiv 2$ (mod 7) $\rightarrow x \equiv 0$ (mod 7)
    Then the code of the year in which was born is 0.
    1975 is the only possible year.

Afterword

How did you do? I hope you learned the mathematics of the Gregorian calendar to the point where you can do all of the tasks mentally, as I do. You may even be faster than I am, especially if you're young. At this moment, I'm 64 years old.

I sincerely hope that you have benefited from this book. My goal was to stimulate minds and promote analytical thinking. If that was the result for you, then I have succeeded. I hope you will use the skills developed to amuse your friends and family. This is my first and last book and I want it to be a catalyst of change.

It's up to you to tell your friends and family how you can do such mental tricks, especially if you have learned enough to do these tasks mentally, as I have. People ask me how I do them but frankly, I don't have the time to tell them! That is why I wrote this book. Time flies. Life is short. I wrote it to pass along the knowledge that I have learned to do tasks like The Birthday Problem. Thus, I expanded my classroom to the world. Mission accomplished.

I believe that learning such mathematics changes the physiology of the brain. Brain function improves when it is exercised with problems such as The Birthday Problem.

The human brain was designed for this kind of analytical thinking. That is what separates us from the *other* animals on this planet. Have fun and I'll see you in *time*.

David Braverman

December 6, 2009

Index

A

absolute value, 14–17
algorithm for day of the week, 11, 66–69, 79, 86, 97, 103–11
algorithm for year code, 11

C

centennial year, 7, 13
century codes, 9, 12, 21, 51–52, 61, 90

D

days of the week, 11, 20–21, 61

E

equations in modulo 7, 41, 62
equivalences in modulo 7, 20

G

Gregorian calendar, 13
Gregory XIII (pope), 13

J

Julian calendar, 13

L

leap day, 13
leap year, 7, 11, 13, 45
leap year + 1, 81

M

minuend, 20
modulo 7, 20
Modulo 7 Clock, The, 10, 20, 25
month codes, 12, 21
multiples of 4, 20
multiples of 7, 7, 20, 23, 79–80

N

negative equivalences, 20

R

reducing in modulo 7, 20

S

signed numbers, 14
subtrahend, 20

www.ingramcontent.com/pod-product-compliance
Lightning Source LLC
Chambersburg PA
CBHW021954170526
45157CB00003B/981